WJEC Chemistry for AS Level

Revision Workbook

Rhodri Thomas

Lindsay Bromley

Published in 2021 by Illuminate Publishing Limited, an imprint of Hodder Education, an Hachette UK Company, Carmelite House, 50 Victoria Embankment, London EC4Y 0DZ

Orders: please contact Hachette UK Distribution, Hely Hutchinson Centre, Milton Road, Didcot, Oxfordshire, OX11 7HH. Telephone: +44 (0)1235 827827. Email: education@hachette.co.uk. Lines are open from 9 a.m. to 5 p.m., Monday to Friday. You can also order through our website: www.hoddereducation.co.uk

British Library Cataloguing in Publication Data

A catalogue record for this book is available from the British Library

ISBN 978-1-912820-66-5

Printed by Ashford Colour Press Ltd

Impression 4
Year 2024

Hachette UK's policy is to use papers that are natural, renewable and recyclable products and made from wood grown in well-managed forests and other controlled sources. The logging and manufacturing processes are expected to conform to the environment regulations of the country of origin.

Publisher: Eve Thould

Editor: Geoff Tuttle

Design and layout: Nigel Harriss

Cover image: Shutterstock.com/Andrey Armyagov

Picture Credits

Illustrations © Illuminate Publishing

Contents

Introduction

PRACTICE QUESTIONS

Unit 1: The Language of Chemistry, Structure of Matter and Simple Reactions

Unit 2: Energy, Rate and Chemistry of Carbon Compounds

PRACTICE EXAM PAPERS

ANSWERS

How to use this book

What this book contains

The AS Chemistry course is made up of two units each assessed with a separate exam. Each unit contributes an equal amount towards the AS course and together they contribute 40% of the A Level course.

Unit 1: The Language of Chemistry, Structure of Matter and Simple Reactions

Unit 2: Energy, Rate and Chemistry of Carbon Compounds

Each unit is divided into areas of study, with seven in Unit 1 (Formulae and equations; Basic ideas about atoms; Chemical calculations; Bonding; Solid structures; The periodic table; Simple equilibria and acid-base reactions) and eight in Unit 2 (Thermochemistry; Rates of reaction; The wider impact of chemistry; Organic compounds; Hydrocarbons; Halogenoalkanes; Alcohols and carboxylic acids; Instrumental analysis). The mathematical content covered in the 'chemical calculations' area of study is considered to be part of all units, so you need to be familiar with this when sitting Unit 2 and the A Level units.

Each area of study has its own section in this book and there you will find:

- A **concept map**, which displays how the different concepts within the area of study are related to one another and to other areas of study.

- A set of **graded questions**, with space for your answers, which are designed to test the content of the area of study in a way which is similar to the examination.

- A **question and answers** section containing some exam-style questions, with marked answers from two students who produce answers of different standard, together with a MARKER COMMENTARY.

The next section comprises two **practice papers** – one for each component. The final section consists of **answers** to all the graded questions and the practice papers.

How to use this book effectively

This book can be used for revision, in which case you should work your way gradually through the questions as you revise each area of study. Alternatively, it can be used regularly as you are learning AS Chemistry, with each set of questions providing an end-of-area check of your work. Your teacher might even like to use it as a homework book because it contains sets of end-of-area questions as well as the practice papers.

As you revise your work in preparation for your exam, you need to ensure that you are processing the information in the area of study. Reading over the content of your notes or textbooks will simply send you to sleep. Only a small fraction of the information will be retained, especially if you are reading over notes for an extended period without switching activities or having a break. One technique that can help is producing summaries of notes in a different format as you work through a unit of study. Even so, you will be preparing yourself only for exam questions in which you reproduce learned material. These so-called Assessment Objective 1 (AO1) questions form only 35% of the AS exams. The higher-level skills required for AO2 and AO3 require different revision techniques. See pages 5 to 7 for further information on assessment objectives.

You will find that an essential component of revision is answering exam-type questions. As in many fields, it is practice that makes perfect in chemistry and you should practise as many past papers and questions within this book as you can. This will allow you to identify the essential details needed when attempting different question types.

When completing the questions, you will, inevitably, come across questions that seem difficult. These questions will allow you to identify which areas you need to focus on during your revision. Should you not be able to answer these questions, you should revisit your notes/textbook. Should the question still leave you stumped, have a look at the answers in the back of the book. If the answer seems unintelligible, it is time to ask your teacher or your fellow students for an explanation of how and why the answer is correct. Your teacher may also point out that different answers are creditworthy, especially in discussion or suggestion questions.

So, answering the questions within these covers and analysing the model answers will do more for your exam preparation than just reading notes or making revision timetables (however colourful)!

Assessment objectives

You need to demonstrate expertise in answering examination questions in different ways. Chemistry requires you to explain your ideas in words but also in calculations. At least 20% of marks across all units will be based on calculations that require mathematics equivalent to a good GCSE grade. You have performed a set of experiments throughout your AS studies and questions can be based on these as well, contributing at least 15% of the marks across the AS units.

All questions are classified as one of three types. These categories are called assessment objectives and are referred to as AO1, AO2 and AO3:

Assessment Objective 1 (AO1)

AO1 questions are ones in which you need to:

demonstrate knowledge and understanding of scientific ideas, processes, techniques and procedures

This objective accounts for 35% of the marks in the two exam papers.

The official definition given in bold (**demonstrate ... procedures**) appears more complex than it is. Essentially, these are the marks that can be obtained without too much thinking. This category covers:

- Recall of terms and their meanings from the specification
- Describing experiments from the **specified practical tasks**
- Recalling the observations expected in familiar chemical reactions and tests
- Recalling and inserting appropriate data into simple equations with no rearrangement.

In general, no judgement or new thinking is required. One can learn an observation or term without fully understanding it, and so the standard of answers given to AO1 questions generally reflects the degree of revision a student has undertaken. These questions are generally straightforward, and you should aim to get a very high proportion of these marks if you want to achieve a high grade.

Examples of AO1 questions

1. **Give a test that can distinguish between cyclohexane and cyclohexene, giving relevant observation(s).** [3]

Good answer (full marks 3/3): Test both compounds with bromine water. The cyclohexene would turn the bromine water from orange to colourless whilst the cyclohexane would leave the bromine water orange.

Poor answer (1/3): Add bromine water and cyclohexene would go colourless.

The poor answer lacks detail in the observation for cyclohexene. Any test to distinguish compounds should refer to the observations for both compounds.

2. **Give the meaning of the term *relative atomic mass*.** [2]

Good answer (2/2): This is the average mass of an atom compared to 1/12th the mass of an atom of carbon-12.

Poor answer (0/2): This is the mass of an atom compared to 1/12th the mass of carbon.

The poor answer is missing 'average'. The 'compared to 1/12th the mass of carbon-12' is present in all relative mass definitions and getting two errors in this (missing 'atom' and '12') is troubling.

3. **State the atom economy for the production of ethanoic acid in the reaction below.** [1]

$$CH_3OH + CO \rightarrow CH_3COOH$$

Good answer (1/1): 100%

Poor answer (0/1): Atom economy = $(12+3+12+16+1) \div (12+3+16+1+12+16) = 73.3\%$

This question is a 'state' question so it shouldn't need a calculation. The good answer identifies 100% as the atom economy for any reaction with only one product. The poor answer loses the mark for an error in the calculation, but also loses time as it involves a calculation.

Assessment Objective 2 (AO2)

AO2 questions are ones in which you need to:

apply knowledge and understanding of scientific ideas, processes, techniques and procedures:
- **in a theoretical context**
- **in a practical context**
- **when handling qualitative data**
- **when handling quantitative data.**

These questions account for 45% of the marks in the two exam papers.

Here, the key words are 'apply knowledge and understanding'. In AO1, demonstrating knowledge and understanding is enough but in AO2 it needs to be applied to specific examples or contexts. These contexts can be similar to, or even the same as, some that you have performed experimentally, or they can be totally new to you but with the experimental method described. Some contexts can be theoretical rather than real.

The tasks require you to apply knowledge using quantitative data, which includes measurements, numbers and calculations, and qualitative data which does not. There will be a balance of questions requiring you to calculate the result and those asking for explanations in words. This is the largest assessment objective with 45% of the marks and you should expect about half of these to be calculations.

Sometimes the application of knowledge in AO2 can also include analysis of data even though 'analyse' also appears in AO3 (see page 7). An AO2 analysis question will usually have more straightforward data and give more clear instructions and guidance. if you are told what type of analysis to carry out, these will be AO2 skills. If the question is more open-ended and you must choose the analysis methods yourself, the question will be classified as AO3.

Examples of AO2 questions

1. **Calculate the volume, in dm^3, of carbon dioxide released by thermal decomposition of 16.4 g of magnesium carbonate, $MgCO_3$, at a pressure of 0.75 atm and a temperature of 700 °C.** [3]

Good answer (3/3): moles of CO_2 = $MgCO_3$ = $^{16.20}/_{84.3}$ $V = nRT / p = 20.5$ dm^3

Poor answer (1/3): moles of CO_2 = $^{16.2}/_{84.3}$ = 0.192 $V = 0.192 \times 8.31 \times 700 / 0.75 = 1489$ dm^3

The errors here are not to interconvert the units in the ideal gas equation (temperature and pressure) and not to change the final answer from m^3 to dm^3.

2. **Explain why the bond angles in the molecules NH_3 and H_2O are different.** [3]

Good answer (3/3): Both molecules have four electron pairs so form shapes based on a tetrahedron: NH_3 has three bond pairs and one lone pair, whilst H_2O has two bond pairs and two lone pairs. Lone pairs repel more than bonded pairs, so the bond angle in H_2O (V-shaped) will be smaller than NH_3 (pyramidal shape).

Poor answer (0/3): They have different angles because they have different shapes. NH_3 has three bond pairs while H_2O has two bond pairs. More bond pairs mean more repulsion so a smaller bond angle for NH_3.

The poor answer recognises that this is VSEPR as it is a question about molecular shapes but is unable to apply the principles to these molecules and totally misses the key idea of lone pairs.

3. **The first ionisation energy of helium is 2370 kJ mol^{-1}. Calculate the wavelength, in nm, of light that corresponds to this energy.** [3]

Good answer (3/3): Energy per atom = $^{2370 \times 1000}/_{N_A}$ = 3.94×10^{-18} J

$E = hf$ and $c = f\lambda$ so $\lambda = ^{hc}/_E = 50.5$ nm

Poor answer (1/3): $E = hfN_A$ so $f = ^h/_{(E \times N_A)}$ = 5.94×10^{12} Hz   wavelength = $c/f = 5 \times 10^4$ nm

This is AO2 as it requires rearranging and combining equations. The poor answer misses the conversion from kJ and gives the final answer to one significant figure which is over-truncation. The rearrangement of equations is correct and gains the only mark.

Assessment Objective 3 (AO3)

AO3 questions are ones in which you need to:

> **analyse, interpret and evaluate scientific information, ideas and evidence, including in relation to issues, to:**
> - **make judgements and reach conclusions**
> - **develop and refine practical design and procedures.**

These questions account for 20% of the marks in the two exam papers. These questions are often answered poorly, and many students find them to be amongst the hardest questions on the exam paper. Because you are expected to analyse, interpret and evaluate, you may find that there is more information presented than in other questions. This can be in the form of text, such as a practical method, or in graphs or tables.

These questions require you to use higher order skills such as analysis, interpretation and evaluation. They can sometimes appear less specific than AO1 and AO2 as you may need to select your own method, choose information from a selection or combine ideas from different parts of the course. They are divided into two types – forming judgements or conclusions and developing and refining practical work.

Questions requiring you to make judgements and reach conclusions are the more common of the two types. They may require you to choose the most appropriate data from a selection or to choose a method to perform a calculation where there are different options. These may guide you to select the most precise or most efficient method, which forces you to evaluate the different methods possible. They may provide more information than is needed and allow you to pick out what is relevant. In some questions you are given a statement and have to determine whether or not (or to what extent) it is true.

The remaining AO3 questions focus on developing and refining practical methods. These question types may give you a method to improve or to explain why particular aspects of a method are used. These may rely on your understanding of practical tasks you have done as part of the course, so you should always review these and think through the reasons for each step.

Examples of AO3 questions

1. **In this experiment you have measured the volume of acid needed for reaction with the sodium hydroxide in the aqueous mixture, and the volume needed for reaction with the sodium hydrogencarbonate. State and explain which set of data is most accurate.** [3]

	1	2	3	4
Volume of NaOH (aq) / cm^3	26.30	26.25	26.35	26.30
Volume of NaHCO$_3$ (aq) / cm^3	12.40	11.90	12.10	11.85

Good answer (3/3): The data for sodium hydroxide appear most accurate. The repeat readings are much closer to each other than the case for sodium hydrogencarbonate. The volume measured is larger than for NaHCO$_3$ so the percentage error is smaller. The volume of NaHCO$_3$ depends on the measurement of two endpoints, while the one for NaOH only requires one so NaOH is less likely to have errors.

Poor answer (1/3): The values for NaOH are repeatable – all four values are close together. This makes the data more accurate. NaOH is a strong base so its endpoint is better than that for the weak base NaHCO$_3$.

The poor answer correctly identifies NaOH and includes the ideas of repeat readings but doesn't give a full and relevant explanation of this choice.

2. **The M_r of the compound is 353. Use the data provided and your answers to earlier parts of the question to find the formula of the mineral.** [2]

Discussion: This is an AO3 question and an example of coming to a conclusion based on a range of data. Earlier parts of the question are likely to involve identification of the ions present in the mineral, the amounts of each and any other species such as water of crystallisation. Combining these together to come to a conclusion makes this part an AO3 question.

Preparing for the examinations

The Units 1 and 2 papers are worth 80 marks and last 1½ hours each. Both papers consist of section A worth 10 marks and section B worth 70.

Structure of the examinations

Section A: Between five and eight short questions worth one or two marks each. The questions cover a wide range of topics and are not linked together. This section usually has a majority of AO1 questions, with a few AO2 questions. AO3 questions are rare in section A.

Section B: This consists of a series of structured questions. The parts within each question may be linked by a theme or context, but still cover a range of areas of study. For example, a question on an industrial preparation may include calculations of yield or atom economy, equilibria questions on maximising the amount of product or evaluation of practical methods. In Unit 1 the work on equations and calculations is rarely treated by itself, and it is usually part of a larger question. In Unit 2 the links between the organic chemistry topics mean that these often appear together within the same questions.

Examination content

The exam papers cover the areas of study listed in the specification and in this book. This work builds upon your previous studies, and elements of GCSE study that overlap with AS work may help you in a small number of questions – remember that some ideas at AS replace those at GCSE and a GCSE-style answer will not gain marks in these cases. Examples of these are electronic structures, ionisation energies and areas of bonding. Similarly, your AS will contribute to the A Level papers as these contain synoptic elements where AS and A2 content are combined.

In addition to the theory work you have studied, you are expected to have performed at least 11 pieces of practical work. Most will have followed the methods listed in the chemistry lab book, although your teachers may have chosen slightly different practical tasks to develop your skills. These may be part of the questions asked in the relevant unit papers – at AS you are only assessed on your practical work in the theory papers, although at A Level there will be a practical exam.

Types of question

Part-questions in the exam will vary in length, with the majority worth between 1 and 4 marks and a very few worth more. There will be one 6-mark question on every exam paper than will be labelled 'QER'. This stands for 'Quality of Extended Response' and these questions are marked differently from others on the paper. The examiner will assess how well you have covered the content of the question, but also how well you have structured your answer, the quality of your language and your use of relevant scientific vocabulary. The question itself can be classed as AO1, AO2, AO3 or a mixture of these. It may include several bullet points, and you should ensure that you discuss all of these.

Each part question will have a key command word or phrase and you should familiarise yourselves with these. If one question asks you to describe an observation, another asks you to give a reason and a third asks you to explain, they all need different answers. Giving an explanation to a describe question is unlikely to gain any marks, even if the content you discuss is correct. The most common command words are listed below.

Give: Usually requires a very short factual answer.

Example: Give the observation for barium chloride in a flame test.

Answer: Apple-green flame

State: Just provide a statement without an explanation.

Example: State the patterns in electronegativity of elements in the periodic table.

Answer: Electronegativity increases across a period and decreases down a group.

State, giving a reason: This expands the statement to include some reasoning. It does not require reference to basic principles as would be expected in a 'State and explain' question.

Example: State, giving a reason, which element in group 1 has the greatest ionisation energy.

Answer: Lithium, as the ionisation energies decrease down the group.

State what is meant by (or give the meaning of ...): This requires all the essential details of the meaning of a term.

Example: State what is meant by the term standard enthalpy change of formation.

Answer: The standard enthalpy change of formation is the enthalpy change when 1 mole of a substance in its standard state is formed from its elements in their standard states under standard conditions.

Define: You need to provide a statement which is (almost) identical to the formal definition.

Example: Define pH.

Answer: $pH = -\log[H^+]$

Describe: Provide a brief description but no explanation is required.

Example: Describe the visible atomic emission spectrum of hydrogen.

Answer: A series of bright lines on a dark background. Each series of lines gets closer together at higher energies until they overlap.

Explain: This usually requires reference to the basic principles of the areas of study, such as the behaviour of electrons, atoms or ions, the bonding between them or the energy changes involved.

Example: Explain the visible atomic emission spectrum of hydrogen.

Answer: The energy in an atom is quantised at particular energy levels. The electrons absorb energy and move to higher energy levels and when they return to lower energy levels, they give out this as a particular frequency of light, found using $E = hf$. In the visible spectrum electrons drop down to the $n = 2$ energy level. The lines get closer together until they overlap at higher energies because the energy levels in the atom get closer together at higher energy levels.

Explain/Give the difference/Compare (between two or more things): This is a comparison question. Depending on the nature of the question it can need either:

- Statements regarding both parts, stating what is present and what is not, e.g. 'Alkenes react with bromine water to change it from orange to colourless as they contain C=C double bonds. Alkanes do not react with bromine water as they do not contain C=C double bonds.' The word 'only' is also useful here, e.g. 'Methane has only Van der Waals forces between its molecules but ammonia has hydrogen bonds between its molecules as well.'

- Comparisons which use words like 'more', 'less', comparative terms for comparisons of two species (words that end -er: stronger, weaker, higher, lower) or superlative terms when you have three or more species (words that end -est: longest, shortest), e.g. 'The rate of reaction is greatest with magnesium powder rather than a single lump or piece of magnesium ribbon as the powder has the greatest surface area, so successful collisions occur most frequently for the powder.'

Suggest ... (or suggest a reason ...): Although not a common command word, this can produce some questions that are difficult to answer. These will often be AO3 marks, appearing at the end of a question requiring evaluation skills, e.g. suggest a reason why method A is favoured according the principles of green chemistry.

Answer: Method A has a greater atom economy so there is less waste and requires lower temperatures and pressures so less energy is needed.

Calculate or **determine**: The aim is to obtain the correct answer (along with the correct unit if required by the mark scheme). With this command word, the correct answer will obtain full marks without the workings. However, you are strongly advised to show your working as marks are available for this even if the answer is wrong. Some questions will require the answer given in specific units and these may appear at the end of the line provided for writing the answer e.g. '........................ dm^3'.

Example: calculate the volume of CO_2 at 298 K and 1 atm pressure that would be produced by adding 1.50 g of $CaCO_3$ to excess acid.

Answer: $n(CO_2) = n(CaCO_3) = 1.50/100.1 = 0.0150$ mol $V = 0.0150 \times 24.5 = 0.367$ dm^3

Find: These questions require a specific answer that may be found through reasoning or calculation.

Justify: This is sometimes used in questions which ask you to justify your answer/decision, e.g. 'Is the student correct? Justify your answer'. It is often used in AO3 questions and may require answers which are similar to 'explain' or 'give reasons' questions depending on the context.

Evaluate: You will be required to make a judgement, e.g. whether a statement is correct or incorrect, whether a method is appropriate, whether data are good or a final value is accurate.

Discuss: This can sometimes be a command word in longer questions. It may suggest that there are positive and negative points to be included, and often these questions award marks for your reasons rather than any conclusion reached.

Common exam mistakes

1. **Misuse of key terms**: A significant minority of answers treat the terms atom, ion and molecule as if they mean the same thing – each is different, and you need to use them correctly. The boiling temperature of HCl is governed by the forces between **molecules** but its chemistry is governed by the covalent bond between the **atoms**. When it dissolves it acts as an acid by dissociating to release H⁺ **ions**.

2. **Missing part of the question**: The most common parts of questions that are omitted are those that do not have dotted lines for you to answer on, e.g. adding to graphs or diagrams. Other common missed questions are ones that have an **and** condition in the question itself, e.g. state and explain. In extended questions there may be bullet points indicating areas that should be included, and the answer should reference all of these.

3. **Not reading the question carefully enough**: This often results in answering a different question altogether by misreading the command words, such as giving a description when the question asks for an explanation. If parts of the question are in bold they are essential – a question stating 'give the **formula** of ...' will not award a mark for a name, and a question asking for a **skeletal formula** will not accept any other representation.

   Similar terms may cause problems if not read correctly: if a question is based on ethene then an answer based on ethane will gain no marks. If you have to describe the atomic emission spectrum of hydrogen, a description of the atomic absorption spectrum will not gain marks.

4. **Units, units, units**: Many calculations will involve changing between units. This is particularly the case for temperature (°C ⟶ K), pressure (Pa ⟶ kPa ⟶ atm) and volume (cm³ ⟶ dm³ ⟶ m³). The use of various SI prefixes may also occur for any unit with masses in mg or tonnes, distances in nm or cm and frequencies in GHz or MHz. The data booklet contains all the interconversion factors you need. Always check if you need to give the answer in particular units – these may appear in the question or at the end of the line provided for your answer.

5. **Poor algebra and rearranging**: There are many mathematical formulae and equations used throughout the AS units. In many cases these can be used without rearrangement, but some questions will require you to adapt or combine equations, e.g. to convert wavelength to energy. Some questions will require you to work back to an unknown value from a given answer, e.g. calculating a bond energy using the enthalpy change of a reaction and the energies of the remaining bonds. pH is one of the few places students may have seen log functions, so you may find it easier to learn that $pH = -\log[H^+]$ but also $[H^+] = 10^{-pH}$.

6. **Over-truncation / Significant figures**: The numbers that appear on your calculator screen need to be considered before they are put down as answers. If an answer is over-truncated, i.e. shortened to one, or sometimes two, significant figures, the answer can be treated as incorrect. Truncating 1.40 to 1 is large percentage difference. There are a few cases where one figure is acceptable in the answer, such as in calculating empirical formulae, but these are in the minority. Some students write their calculation methods down and round their values at each stage, and this can build up to give an incorrect answer overall – if possible, keep the number in your calculator memory or on screen. Final answers are only penalised for too many significant figures if the question asks for a specific number or an appropriate number.

7. **Practical work**: Learn your practical methods as well as your theory. There are marks for practical method recall, and understanding of how to improve practical methods, in every unit paper.

8. **Missing comparisons**: If you are comparing, you need to refer to both species either by stating one has something and the other does not or by making a comparison, e.g. the boiling temperature of A is **higher** than B as the forces between the molecules of A are **stronger** than between molecules of B.

Unit 1: The Language of Chemistry, Structure of Matter and Simple Reactions

1.1: Formulae and equations

Topic summary

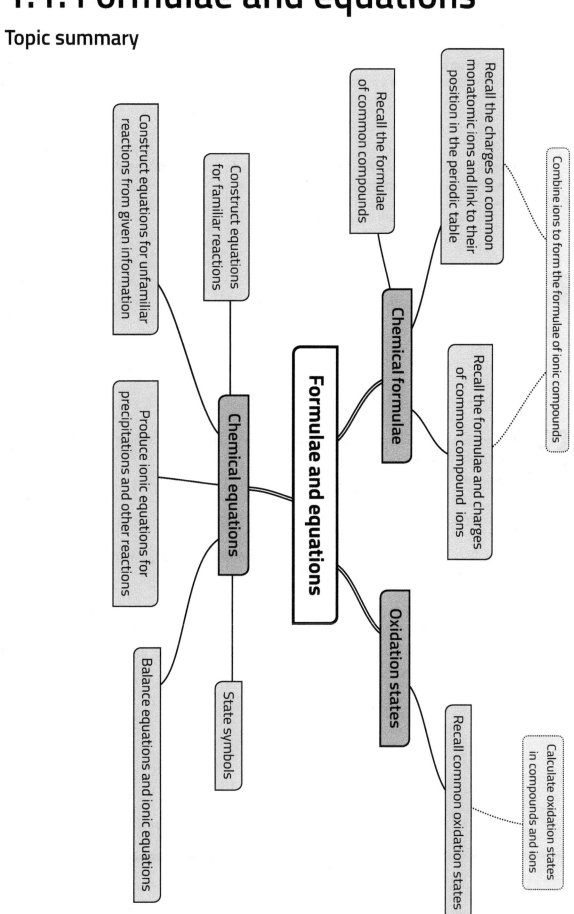

Formulae and equations

Chemical formulae

- Recall the charges on common monatomic ions and link to their position in the periodic table
- Recall the formulae of common compounds
- Recall the formulae and charges of common compound ions
- Combine ions to form the formulae of ionic compounds

Chemical equations

- Construct equations for unfamiliar reactions from given information
- Construct equations for familiar reactions
- Produce ionic equations for precipitations and other reactions
- Balance equations and ionic equations
- State symbols

Oxidation states

- Recall common oxidation states
- Calculate oxidation states in compounds and ions

Q1 (a) Precipitation reactions are often used for identification of unknown ions in solution. Write ionic equations for the reactions occurring in each of the following chemical tests. You should include state symbols.

(i) Barium chloride is used to test for sulfate ions in solution. A white precipitate of barium sulfate is a positive result. [2]

....

(ii) Sodium hydroxide solution is used to test for Fe^{2+} ions in solution. A dark green precipitate of iron(II) hydroxide is a positive result. [2]

....

(b) One method of testing for chlorine gas is to bubble the gas through a solution of sodium bromide. A positive result is an orange solution containing aqueous bromine.

(i) Write an ionic equation for this reaction. [2]

....

(ii) Use oxidation states to show that this is a redox reaction. [2]

....

....

....

Q2 (a) When ethanoic acid is added to copper(II) carbonate, the salt produced is copper(II) ethanoate amongst other products.

Write the equation for this reaction. [2]

....

(b) When any acid is added to a solution of sodium thiosulfate, $Na_2S_2O_3$, the solution turns cloudy as sulfur forms and sulfur dioxide gas is released.

$$2H^+ + S_2O_3^{2-} \longrightarrow S + SO_2 + H_2O$$

Find the oxidation states of sulfur in all the sulfur-containing species in the reaction. [2]

$S_2O_3^{2-}$ ....

S ....

SO_2 ....

Q3   Aluminium oxide reacts slowly with acids but when heated with excess sulfuric acid forms a solution of aluminium sulfate.

(a)   Write an equation for this reaction. [2]

....

(b)   Addition of sodium hydroxide solution to the mixture neutralises the acid followed by the formation of a precipitate of aluminium hydroxide.

(i)   Write an ionic equation for the neutralisation of the acid. Include state symbols. [1]

....

(ii)   Write an ionic equation for the formation of the aluminium hydroxide precipitate. Include state symbols. [1]

....

Q4   'Red lead' is used in anti-corrosive paint and contains mainly the oxide Pb_3O_4. This can be produced by oxidising lead(II) oxide by oxygen gas.

(a)   Write an equation for this reaction. [2]

....

(b)   Use oxidation states to show that the lead has been oxidised in this process. [2]

....

....

....

Unit 1 Practice questions

Question and mock answer analysis

Q&A 1 Phosphoric acid, H_3PO_4, is used to make a range of phosphate salts. These are mixed with other substances to make agricultural fertilisers.

(a) Some of the phosphate ions in fertilisers are trapped as insoluble calcium phosphate when the fertiliser is spread.

    (i) State the formula of the compound calcium phosphate. [1]

    (ii) Write an equation for the formation of calcium phosphate from phosphoric acid, H_3PO_4, and calcium carbonate, $CaCO_3$. [2]

(b) Phosphorous acid, H_3PO_3, is a different acid that decomposes on heating to form phosphoric acid and phosphine, PH_3.

    (i) Write the equation for this reaction. [1]

    (ii) Use oxidation states to show that this is a redox reaction. [2]

Interpretation

At A Level you are expected to recall a range of common ions and deduce less familiar ions from given information. Here you need to recall the calcium ion and work out the formula and charge of the phosphate ion from the formula of the phosphoric acid.

All equations at A Level are symbol equations and are often more challenging than many you have seen previously, so don't be surprised to need numbers greater than 1, 2 or 3.

You can be expected to show oxidation and reduction in different ways, but if the question refers to oxidation states then you will not gain credit for other methods.

Dewi's answer

(a) (i) $Ca_3(PO_4)_2$ ✓

    (ii) $2H_3PO_4 + 3CaCO_3 \longrightarrow Ca_3(PO_4)_2 + 3CO_2 + 3H_2O$ ✓✓

(b) (i) $4H_3PO_3 \longrightarrow 3H_3PO_4 + PH_3$ ✓

    (ii) The oxidation state of oxygen is –2 and hydrogen is +1 on both sides of the equation. At the start the phosphorus is +3 but it changes to –3 in PH_3 and to +5 in H_3PO_4. This shows that both oxidation and reduction have occurred. ✓ ✗

MARKER COMMENTARY

(a) (i) This gains the mark.

    (ii) Dewi has recalled the products of a metal carbonate with an acid correctly from his previous studies. He has also balanced the equation correctly.

(b) (i) The mark here is solely for balancing, as the formulae of all the compounds have been given.

    (ii) The oxidation states given for phosphorus are correct and gain a mark. Dewi states that oxidation and reduction have occurred but does not link this to the oxidation state values so does not gain the second mark.

Overall Dewi gains 5 out of 6 marks as all the content is correct but he does not give enough detail to gain the final mark – references to oxidation and reduction need to link to oxidation states becoming more positive or more negative.

Rebecca's answer

(a) (i) $Ca_3(PO_4)_2$ ✓

(ii) $2H_3PO_4 + 3CaCO_3 \longrightarrow Ca_3(PO_4)_2 + 3CO_2 + 2H_2O$ ✓ X

(b) (i) $4H_3PO_3 \longrightarrow 3H_3PO_4 + PH_3$ ✓

(ii) Hydrogen is reduced as its oxidation state goes from +1 to a mixture of +1 in H_3PO_4 and –1 in PH_3. Phosphorus is +3 in H_3PO_3 and stays as this in PH_3. X X

MARKER COMMENTARY

(a) (i) The correct formula gains the mark.

(ii) Rebecca has remembered that acids form a salt, water and carbon dioxide in their reactions with metal carbonates and so gains a mark for identifying the products. Her equation is not correctly balanced so she cannot gain the second mark.

(b) (i) She has correctly balanced the equation and so she gains 1 mark.

(ii) The oxidation states given for phosphorus and hydrogen at the start are correct; however, those in the products are incorrect. She does not show that the redox reaction includes both reduction and oxidation and so cannot gain the second mark.

Overall Rebecca gains 3 out of 6 marks as she can write correct formulae and balance some equations but cannot use oxidation states for unfamiliar compounds and reactions.

Q&A 2   Silver bromide, AgBr, was used in early photography due to this compound decomposing in the presence of light.

(a) Silver bromide is a precipitate formed when silver nitrate solution is added to a solution containing bromide ions.

(i) Write the ionic equation for this reaction. Include state symbols.   [2]

(ii) Give the colour of the precipitate of silver bromide.   [1]

(b) The decomposition of silver bromide in the presence of light produces the elements silver and bromine.

(i) Write an equation for this process.   [1]

(ii) Show that this is a redox reaction whilst the precipitation reaction in part (a) is not a redox reaction.   [3]

(c) Many unreactive metals such as silver do not react easily with acids.

(i) Silver can react with nitric acid in the presence of air to give silver nitrate, water and nitric oxide, NO. Balance the equation for this reaction.   [1]

$$\text{......... } Ag + \text{......... } HNO_3 \rightarrow \text{......... } AgNO_3 + \text{......... } NO + \text{......... } H_2O$$

(ii) Other unreactive metals only dissolve in aqua regia, a mixture of concentrated nitric and hydrochloric acid that react to produce products including Cl_2 and NOCl. Complete the equation for this process.   [2]

$$\text{......... } HNO_3 + \text{......... } HCl \rightarrow \text{...}$$

Interpretation

At A Level you are expected to build upon the work you have studied previously, which includes common reactions, such as those of acids and bases and simple chemical tests, and this question requires you to recall these.

All equations at A Level are symbol equations but state symbols are only required where stated, such as in part (a).

You can be expected to show oxidation and reduction in different ways, and in part (b) (ii) you are free to use any appropriate method as the question does not limit you. In most cases oxidation states are the most straightforward but gain or loss of electrons is also acceptable in this example.

Dewi's answer

(a) (i) $Ag^+ (aq) + Br^- (aq) \longrightarrow AgBr (s)$ ✓✓

    (ii) cream ✓

(b) (i) $AgBr (s) \longrightarrow Ag (s) + Br_2 (aq)$ ✓

    (ii) For a redox reaction to occur, electrons must be transferred from one species to another. The one that loses electrons is oxidised, and the one that gains electrons is reduced. In the decomposition reaction the silver is reduced as the reaction occurring is $Ag^+ + e^- \longrightarrow Ag$. The bromine is oxidised as it loses electrons in $2Br^- \longrightarrow Br_2 + 2e^-$.

    In the precipitation reaction the silver starts as silver ions and stays as silver ions – they have not lost or gained electrons, so it is not a redox reaction. The same is true for the bromide ions. ✓✓✓

(c) (i) $3Ag + 3HNO_3 \longrightarrow 3AgNO_3 + 1NO + 2H_2O$ ✓

    (ii) $1HNO_3 + 3HCl \longrightarrow NOCl + Cl_2 + 2H_2O$ ✓✓

MARKER COMMENTARY

(a) (i) The equation is correct, and the state symbols are also correct so this gains 2 marks.

(ii) Dewi has correctly recalled the colour of this precipitate from previous work and will also see it later in this unit.

(b) (i) State symbols are not needed here so the error in these is ignored for the mark – the substances in the equation are correct.

(ii) Dewi has chosen to use electron transfer to answer this question, which is an appropriate method. He gains a mark for explaining what is meant by a redox reaction in terms of electrons. He then applied this idea correctly to each of the two reactions and gains a mark for each of these for a total of 3 marks.

(c) These equations are more challenging than those encountered at GCSE. Dewi has balanced the equation in (i) correctly. Once he identified the additional product in part (ii) as water by the need to balance the hydrogen atoms, he balances the equation correctly.

Overall Dewi gains 10 out of 10 marks as he recalls all the factual content correctly and can apply these ideas to the reactions of silver bromide discussed.

Rebecca's answer

(a) (i) $Ag^+ (aq) + Br^- (aq) \longrightarrow AgBr (aq)$ ✓✗

    (ii) cream ✓

(b) (i) $AgBr \longrightarrow Ag + Br$ ✗

    (ii) In a redox reaction oxidation states must change showing that oxidation and reduction have happened – oxidation states become more positive during oxidation. In the precipitation the oxidation states don't change so it isn't a redox reaction. In the decomposition reaction the silver's oxidation state decreases to 0 so it is reduced, while the bromide ions are oxidised with the oxidation states becoming less negative (−1 to 0). ✓✗✓

(c) (i) $3Ag + 3HNO_3 \longrightarrow 3AgNO_3 + 1NO + 2H_2O$ ✓

    (ii) $HNO_3 + HCl \longrightarrow NOCl + Cl_2 + H_2$ ✗✗

MARKER COMMENTARY

(a) (i) The equation is correct but the state symbol for the precipitate should be (s) so this gains only 1 mark.

(ii) Rebecca has correctly recalled the colour of this precipitate.

(b) (i) Rebecca has not written the correct formula for bromine as an element (Br_2) so the equation is incorrect.

(ii) Rebecca has chosen to use oxidation states to answer this question, which is another appropriate method. Her discussion of what is meant by a redox reaction is clear and gains a mark.

When discussing the precipitation reaction, she correctly states that oxidation states don't change, but doesn't state what they are, so this isn't enough for the second mark. Her explanation of the decomposition reaction is better and uses correct oxidation states for most substances to gain the third marking point.

(c) Rebecca can balance the first equation correctly. She realises there must be a hydrogen-containing product but suggests this is H_2 – it is impossible to balance this equation so she loses the second mark.

Overall Rebecca gains 5 out of 10 marks as she has many correct ideas but careless errors on formulae and state symbols have reduced her marks. Her work on oxidation states is correct but not enough to gain all the marks available.

1.2: Basic ideas about atoms

Topic summary

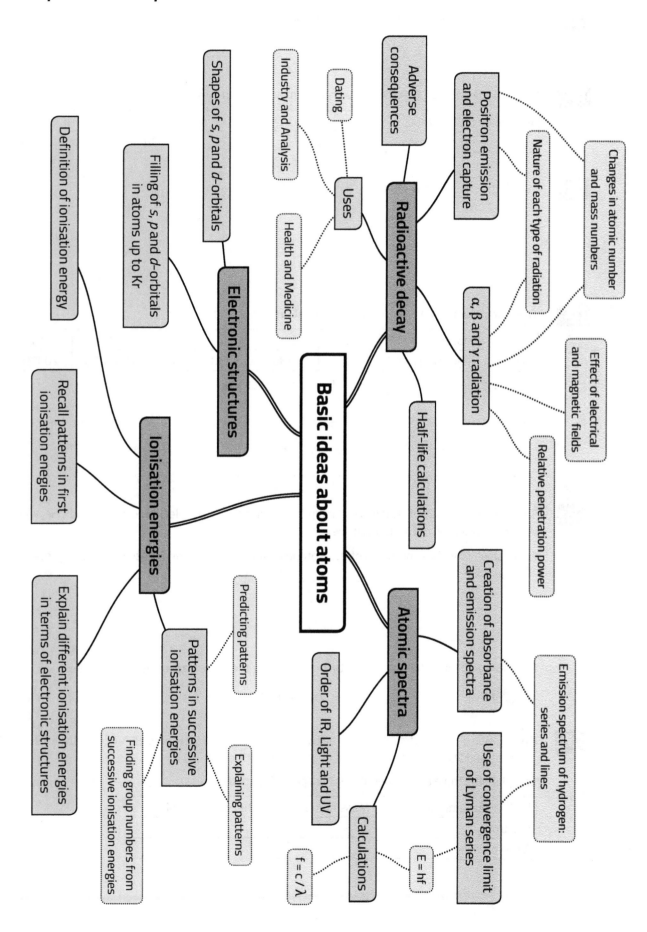

The mind map "Basic ideas about atoms" includes the following branches and nodes:

Electronic structures
- Shapes of s, p and d-orbitals
- Filling of s, p and d-orbitals in atoms up to Kr

Radioactive decay
- Adverse consequences
- Uses
  - Industry and Analysis
  - Dating
  - Health and Medicine
- Positron emission and electron capture
- Nature of each type of radiation
- α, β and γ radiation
  - Changes in atomic number and mass numbers
  - Effect of electrical and magnetic fields
  - Relative penetration power
- Half-life calculations

Ionisation energies
- Definition of ionisation energy
- Recall patterns in first ionisation energies
- Explain different ionisation energies in terms of electronic structures
- Patterns in successive ionisation energies
  - Predicting patterns
  - Explaining patterns
  - Finding group numbers from successive ionisation energies

Atomic spectra
- Order of IR, Light and UV
- Creation of absorbance and emission spectra
- Emission spectrum of hydrogen: series and lines
- Use of convergence limit of Lyman series
- Calculations
  - $E = hf$
  - $f = c/\lambda$

Q1 Identify an element that has a half-filled p sub-shell. [1]

....

Q2 Write the electron arrangement of the oxide ion, O^{2-}. [1]

....

Q3 Show the electron arrangement of a chromium atom as arrows in boxes. [1]

Neon core		$3s$			$3p$					$3d$				$4s$

Q4 Sketch the shape of a p-orbital. [1]

Q5 The first six ionisation energies of an element are shown in the table:

Electron removed	1st	2nd	3rd	4th	5th	6th
Ionisation energy / kJ mol⁻¹	786	1580	3230	4360	16000	20000

Find the group number of this element, giving your reason(s). [2]

....

....

....

Q6 Selenium has a range of isotopes including ^{72}Se, which decays by electron capture, and ^{82}Se, which decays by emission of two ß⁻ particles.

(a) Suggest why the mass numbers of the isotopes do not change during these radioactive processes. [1]

....

....

(b) Give the symbols of the two elements formed during these radioactive decay processes. [1]

Decay of ^{72}Se .... *Decay of ^{82}Se* ....

Q7 One isotope of curium, ^{233}Cm, decays by two different radioactive routes: 80% decays by emission of a positron, and 20% decays by emission of an alpha particle.

Complete the table below, identifying the species formed. [2]

Radioactive emission	Mass number	Chemical symbol
Positron		
Alpha particle		

Q8 Radioactive decay can be classed as α, β⁻, γ or positron emission.

(a) Complete the table below noting the changes in atomic number and mass number when each type of radiation is emitted. [2]

Type of radioactivity	α	β⁻	γ	positron
Change in atomic number	-2			
Change in mass number	-4			

(b) Complete the diagram below showing the effects of passing each radiation type through an electrical field. The path of the α-particle has been completed for you. [2]

β

γ

α

positron

(c) Polonium-210 is a radioactive isotope that is an α-emitter. It has been suggested that this material has been used by spies to poison their targets. Suggest why it is possible to carry polonium-210 compounds safely as solutions in glass bottles even though the isotope is deadly when ingested. [2]

....

....

....

(d) Give one use of radioisotopes in medicine. [1]

....

Q9 Naturally occurring uranium contains mainly ^{238}U with small amounts of ^{234}U and ^{235}U.

(a) Complete the table to show the numbers of protons, neutrons and electrons in each isotope. [2]

	^{234}U	^{235}U	^{238}U
Number of protons			
Number of neutrons			
Number of electrons			

(b) The isotope ^{234}U decays by alpha emission with a half-life of 2.45×10^5 years.

(i) Identify the isotope formed by alpha emission from ^{234}U. [1]

Symbol .... Mass number .... Atomic number ....

(ii) A sample of uranium contains 5.0×10^{-3} % of ^{234}U. Calculate the percentage of ^{234}U remaining in the sample after 9.8×10^{5} years. [3]
You may assume that the decay of the other isotopes is negligible over this time.

^{234}U remaining .... %

Q10 Almost all the thorium present on Earth is thorium-232. Part of the radioactive decay series of ^{232}Th is shown:

^{232}Th $\xrightarrow{\hspace{2cm}}$ ^{228}Ra $\xrightarrow{\hspace{2cm}}$ ^{228}Ac $\xrightarrow{\hspace{2cm}}$ ^{228}Th

$t_{\frac{1}{2}} = 1.4 \times 10^{10}$ year $\qquad$ $t_{\frac{1}{2}} = 5.8$ year $\qquad$ $t_{\frac{1}{2}} = 6.13$ hr

(a) Indicate the type of radioactive decay occurring in each stage by writing α, β^{-} or β^{+} above each arrow. [1]

(b) Suggest why the isotopes ^{228}Ra and ^{228}Ac do not build up in significant amounts in rocks containing ^{232}Th. [2]

....

....

....

Q11 Bismuth-212 can decay to thallium-208 with a half-life of 60 minutes.

(a) This change requires the emission of an α-particle.

(i) State what is meant by an α-particle. [1]

....

....

(ii) Give one reason why radioactivity is harmful. [1]

....

....

(iii) A student states that the fact that this decay process of ^{212}Bi produces ^{208}Tl as the only daughter nucleus shows that α-emission is the only radioactivity produced. State and explain whether the student is correct. [2]

....

....

(b) 2.2 g of bismuth-212 is left for 2 hours. Calculate the mass of bismuth-212 remaining. [2]

Mass of ^{212}Bi remaining ....

Q12 The six successive ionisation energies of carbon are given in the table:

Electron removed	1st	2nd	3rd	4th	5th	6th
Ionisation energy / kJ mol^{-1}	1090	2350	4610	6220	37800	47000

(a) Explain why the fifth and sixth ionisation energies of carbon are much greater than the first four ionisation energies. [2]

....

....

....

(b) The sixth ionisation energy of nitrogen is 71000 kJ mol^{-1}.

Explain why this is greater than the sixth ionisation energy of carbon. [2]

....

....

....

(c) The sixth ionisation energy of oxygen is 17900 kJ mol^{-1}.

Explain why this is lower than the sixth ionisation energy of carbon. [2]

....

....

....

Q13 All the energy levels of a hydrogen atom are shown in the diagram:

(a) Using an arrow, indicate on the diagram the electron transition that corresponds to the first line of the Lyman series of the absorption spectrum. Label this 1. [1]

(b) Using an arrow, indicate on the diagram the electron transition that corresponds to the second line of the visible emission spectrum of hydrogen. Label this 2. [1]

(c) Using an arrow, indicate on the diagram the electron transition that corresponds to the ionisation of the atom. Label this 3. [1]

(d) The ionisation energy of hydrogen in 1312 kJ mol^{-1}. Calculate the wavelength of the electromagnetic wave that has this energy. [3]

.... nm

Q14 Sodium exists almost exclusively as one isotope, sodium-23 although the radioactive isotopes sodium-22 and sodium-24 also exist.

(a) Sodium-22 decays by positron emission and sodium-24 decays by β^- emission.

Identify the element formed in each case. [1]

Decay of sodium-22 forms .... Decay of sodium-24 forms ....

(b) Sodium forms Na$^+$ ions when it reacts with many other elements.

Use arrows in boxes to show the electron arrangement of an Na$^+$ ion. [1]

(c) Potassium lies below sodium in group 1. State and explain which of these two elements will have the higher ionisation energy. [2]

....

....

....

(d) Sodium has an atomic number of 11, preceded by neon (atomic number 10) and followed by magnesium (atomic number 12). Place these elements in order of increasing ionisation energy, explaining the patterns in the ionisation energies. [3]

Lowest .... *Highest*

....

....

....

....

Q15 The first ionisation energies of some of the first twelve elements are shown in the table.

Element	H	He	Li	Be	B	C	N	O	F	Ne	Na	Mg
First ionisation energy / kJ mol^{-1}	1312	2370	520	900	800	1090	1400		1680	2080		736

(a) Write an equation for the first ionisation of lithium. [1]

....

(b) Explain why the first ionisation of helium is greater than that of hydrogen. [2]

....

....

....

(c) Explain why the first ionisation energy of lithium is much lower than that of helium. [2]

....

....

....

(d) Suggest a value for the first ionisation energy of sodium. Give reasons for your answer. [2]

....

....

....

(e) Suggest a value for the first ionisation energy of oxygen. Explain why oxygen would have this ionisation energy. [3]

....

....

....

Q16 (a) Sketch and explain the pattern of successive ionisation energies for an aluminium atom. [4]

(b) The first ionisation energy of magnesium is greater than the first ionisation energy of aluminium. Explain why this is the case. [2]

(c) The final ionisation energy of magnesium is lower than the final ionisation energy of aluminium. Explain why this is the case. [2]

Question and mock answer analysis

Q&A 1 Tellurium is one of the rarest stable elements present on Earth. It often forms tellurides containing Te^{2-} ions.

(a) Give the number of protons and electrons present in a telluride anion. [1]

(b) Tellurium forms a range of radioactive isotopes including ^{107}Te, ^{117}Te and ^{127}Te.

    (i) The isotope ^{107}Te can decay by emission of an α-particle or a positron.
Identify the isotopes produced from ^{107}Te by these two decay methods. [2]

    (ii) The isotope ^{127}Te releases a β-particle when it decays. State what is meant by a β-particle. [1]

    (iii) The half-life of the isotope ^{117}Te is 62 minutes. A sample of 80 mg of the isotope decays to leave 5 mg of the isotope after t minutes. Calculate the value of t. [2]

(c) Explain why radioactivity is harmful to health. [2]

Interpretation

In this unit, you build upon your GCSE work, and are expected to recall your work on radioactive decay and atomic structure.

In this question you should apply your knowledge of the formation of ions and radioactive decay to find the numbers of protons, neutrons and electrons in various species – it is important to remember that if the number of protons changes then the element symbol must also change.

You are only expected to use full half-lives in your calculations. Remember time can be given in a range of units and the answer must match.

Dewi's answer

(a) 52 protons and 54 electrons ✓

(b) (i) Alpha decay produces ^{103}Sn and positron emission produces ^{107}I ✓ ✗

    (ii) β-particles are fast-moving electrons. ✓

    (iii) 80 mg ÷ 5 mg = 16 = 2^4 so four half-lives
Time $t = 62 \times 4 = 248$ minutes ✓ ✓

(c) Radioactive particles are high-energy and can damage biological molecules. In large doses this can denature enough molecules to cause death, but smaller doses can damage DNA leading to mutations and then cancerous tumours. ✓ ✓

MARKER COMMENTARY

(a) The correct numbers of protons and electrons.

(b) (i) The mass numbers in both cases are correct; however, Dewi has increased the atomic number by 1 when the positron is emitted rather than decreased it by 1 and so does not gain the second mark.

    (ii) The definition here is correct and gains 1 mark.

    (iii) This method is correct and gains both marks.

(c) This answer is complete and gains 2 marks.

Overall Dewi gains 7 out of 8 marks as he recalls the factual content in detail and can apply ideas to find the numbers of protons, neutrons and electrons in many species.

Rebecca's answer

(a) 52 protons and 52 electrons ✗

(b) (i) The α radioactivity produces tin and the positron produces antimony. ✗ ✗

(ii) β-particles are electrons. ✗

(iii) 80mg → 40mg → 20mg → 10mg → 5mg
Four half-lives so the time is 62 × 4 = 248 minutes ✓✓

(c) Radioactivity can break bonds in DNA and cause mutations. This can then lead to the development of cancers. ✓✓

MARKER COMMENTARY

(a) Rebecca writes the same number of protons and electrons in an ion which is a common error.

(b) (i) The question asks for the isotopes produced, but no mass numbers are given so it is awarded 0 marks.

(ii) Although β-particles are electrons there is no reference to their speed or energy, so no marks are awarded.

(iii) This is the most common correct method seen for calculations using half-lives and gains 2 marks.

(c) This is sufficient for both marks as she discusses damage to DNA and its effect.

Overall Rebecca gains 4 out of 8 marks. She has not read the question carefully enough, leading to simple errors such as giving an answer for an atom in place of an ion and identifying an element and not a specific isotope.

Q&A 2 This question refers only to the elements in the section of the periodic table shown:

Li	Be	B	C	N	O
Na	Mg	Al	Si	P	S

(a) Use arrows in boxes to show the electron arrangement for nitrogen. [1]

(b) Identify the element with the lowest first ionisation energy. Give reasons for your answer. [2]

(c) State which of the elements Be and B will have the greatest ionisation energy. Explain your answer. [2]

(d) Magnesium usually forms Mg^{2+} in its compounds:

$$Mg\,(g) \rightarrow Mg^{2+}\,(g) + 2\,e^-$$

The energy change for this process is 2187 kJ mol^{-1}. Suggest a value for the second ionisation energy of magnesium, giving a reason for your answer. [2]

Interpretation

These questions include some which require explanations and some which ask for reasons. 'Explain' questions require an answer which refers to the basic principles or ideas, in this case the factors which govern the strength of the attraction between nucleus and electron. Questions requiring reasons do not need the same depth of answer, and in this case discussion of patterns in ionisation energies is enough.

Dewi's answer

(a)

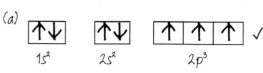

$1s^2$ $2s^2$ $2p^3$ ✓

(b) Ionisation energies increase across a period, so group 1 elements have the lowest ionisation energies. Ionisation energies decrease down the group, so the lowest ionisation energy is the one for sodium. ✓✓

(c) Be will have the greatest ionisation energy because ionisation energies decrease slightly from group 2 to 3. ✓✗

(d) I think the second ionisation of magnesium will be about 1300 kJ mol⁻¹. I have chosen this answer as the second ionisation of magnesium will be slightly larger than the first, so it will be a bit more than half 2187 kJ mol⁻¹. ✓✓

MARKER COMMENTARY

(a) This is the correct electron arrangement.

(b) The marks here are for the reasons. The answer identifies the patterns across a period and down a group, and so gains both marks.

(c) The correct element is identified; however, the candidate gives a reason rather than an explanation. This answer gains one mark only.

(d) The value suggested is within the acceptable range, and the reason given is correct, so it gains both marks.

Overall Dewi gains 6 out of 7 marks as he can correctly identify the electron arrangements of different atoms and link these to their ionisation energies.

Rebecca's answer

(a)

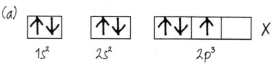

$1s^2$ $2s^2$ $2p^3$ ✗

(b) The general trend in the periodic table is low values in the bottom left and high values in the top left. Sodium will therefore have the lowest value. ✓✓

(c) B will have the greatest ionisation energy because ionisation energies generally increase across the period. ✗✗

(d) I estimate the value to be about 1100 kJ mol⁻¹. I have chosen this answer as the value given is for two electrons, so the value for one will be about half. The second ionisation energy will be slightly higher than the first, and the first ionisation energy in this case will be 1087 kJ mol⁻¹. ✗✓

MARKER COMMENTARY

(a) This is incorrect – the electrons in the p sub-shell should be unpaired.

(b) This is a broad description of the pattern of ionisation energies, and this is just sufficient for the marks.

(c) The element is incorrect. The pattern stated is correct, but it does not apply to the comparison between Be and B which is one of the exceptions.

(d) The pattern described is correct and gains a mark. The value is too low, however, and doesn't gain the second mark.

Overall Rebecca gains 3 out of 7 marks. Her answer shows some awareness of general patterns but lacks the detail needed to gain higher marks.

Unit 1 Practice questions

1.3: Chemical calculations

Topic summary

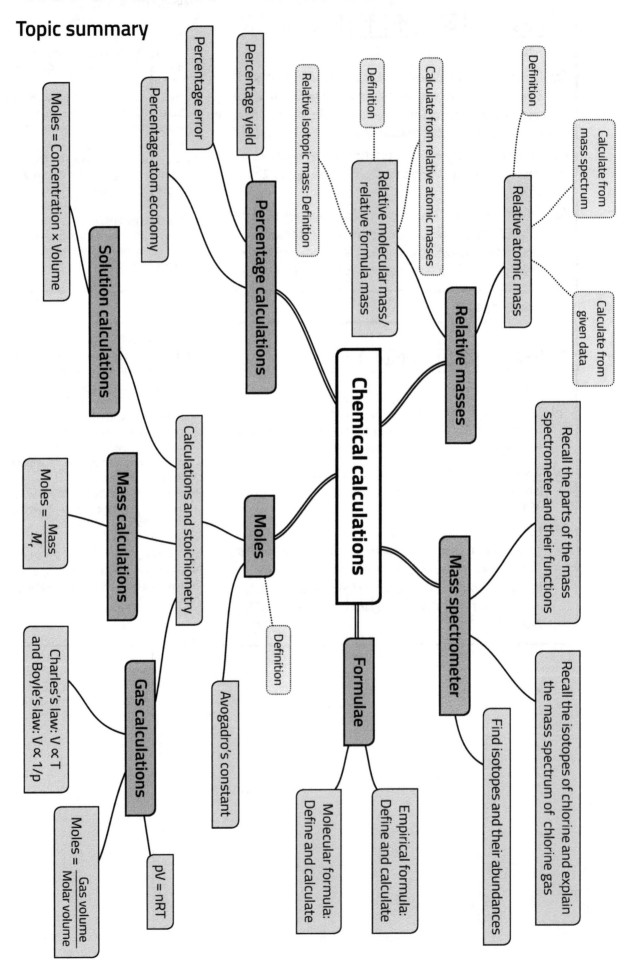

Percentage yield

Percentage error

Percentage atom economy

Solution calculations

Moles = Concentration × Volume

Percentage calculations

Relative Isotopic mass: Definition

Relative molecular mass/ relative formula mass

Definition

Calculate from relative atomic masses

Definition

Relative atomic mass

Calculate from mass spectrum

Calculate from given data

Relative masses

Chemical calculations

Calculations and stoichiometry

Mass calculations

$Moles = \dfrac{Mass}{M_r}$

Moles

Definition

Avogadro's constant

Formulae

Mass spectrometer

Recall the parts of the mass spectrometer and their functions

Recall the isotopes of chlorine and explain the mass spectrum of chlorine gas

Find isotopes and their abundances

Molecular formula: Define and calculate

Empirical formula: Define and calculate

Gas calculations

Charles's law: V ∝ T and Boyle's law: V ∝ 1/p

$Moles = \dfrac{Gas\ volume}{Molar\ volume}$

$pV = nRT$

Q1 State what is meant by the term *mole*. [2]

....

....

Q2 The M_r of an oxide of nitrogen is 108. Find the molecular formula of this oxide. [1]

Molecular formula ....

Q3 Give the meaning of *Avogadro's number*. [1]

....

....

Q4 A sample of $SOCl_2$ has a mass of 2.38 g.

(a) Find the number of molecules in this sample. [2]

Molecules ....

(b) Find the number of atoms in this sample. [1]

Molecules ....

Q5 Give the meaning of the term *empirical formula*. [1]

....

....

Q6 Bromine has two stable isotopes, ^{79}Br and ^{81}Br. Naturally occurring bromine contains 50.69% ^{79}Br. Calculate the relative atomic mass of bromine. Give your answer to four significant figures. [3]

Relative atomic mass ....

Q7 Sodium manganate is a compound of sodium, manganese and oxygen only. It contains 38.8% oxygen and 27.9% sodium by mass.

Find the empirical formula of sodium manganate. [2]

Empirical formula ....

Q8 Sodium carbonate exists as a range of hydrates, $Na_2CO_3.xH_2O$.

(a) One hydrate has a relative formula mass of 286. Find the value of x in this formula. [2]

x = ...

(b) Another hydrate contains 19.8% sodium by mass. Find the value of x in this formula. [2]

x = ...

Q9 Define the term *relative molecular mass*. [2]

....

....

Q10 Pyrogallol contains 57.1% carbon, 38.1% oxygen and 4.8% hydrogen by mass.

(a) Find the empirical formula of pyrogallol. [2]

Empirical formula ....

(b) The relative molecular mass of pyrogallol is approximately 125. Find the molecular formula of pyrogallol. [2]

Molecular formula ....

Q11 Oxalic acid, HOOC–COOH, is present in many fruits and vegetables.

(a) Give the empirical formula of oxalic acid. [1]

....

(b) Oxalic acid can be oxidised to produce carbon dioxide:

$$HOOC–COOH \text{ (s)} + [O] \rightarrow 2CO_2 \text{ (g)} + H_2O \text{ (l)}$$

Calculate the volume of carbon dioxide gas at 1 atm pressure at a temperature of 273 K produced when 6.22 g of oxalic acid is oxidised. [3]

Volume of CO_2 .... cm³

Q12 Xenon is a noble gas. Unlike most noble gases it forms compounds with electronegative elements such as oxygen and fluorine.

(a) The solid XeO_3 decays to release xenon and oxygen gases:

$$2XeO_3 \text{ (s)} \longrightarrow 2Xe \text{ (g)} + 3O_2 \text{ (g)}$$

Calculate the volume of gas produced from 27.0 g of XeO_3 decomposing completely at a temperature of 10 °C. [3]

Volume of xenon .... dm³

(b) A fluoride of xenon reacts with sodium fluoride to form sodium fluoroxenate. It contains 14.0% sodium, 39.8% xenon and 46.2% fluorine by mass. Find the empirical formula of sodium fluoroxenate. [2]

Empirical formula ....

(c) 10.0 g of XeF_2 reacts with an excess of silicon according to the equation shown:

$$2XeF_2 + Si \longrightarrow SiF_4 + 2Xe$$

Calculate the maximum mass of SiF_4 that could be formed. [3]

Mass of SiF_4 .... g

Q13 Ammonia is a very soluble gas. At a temperature of 25 °C a mass of 31.0 g of ammonia can dissolve in 100 cm³ of water.

(a) Calculate the concentration of this solution.

Assume the volume of the solution does not change when ammonia is added. [2]

Concentration .... mol dm³

(b) Calculate the volume of ammonia that dissolves in 100 cm³ of water at a temperature of 25 °C at a pressure of 1 atm. [1]

Volume .... dm³

Q14 Glucose has the molecular formula $C_6H_{12}O_6$.

(a) Give the empirical formula of glucose.   [1]

....

(b) The overall equation for the production of glucose by photosynthesis is:

$$6CO_2 + 6H_2O \rightarrow C_6H_{12}O_6 + 6O_2$$

(i) Calculate the atom economy of this process.   [2]

Atom economy .... %

(ii) Calculate the volume of oxygen gas, measured at 310 K and 1 atm pressure, produced during the formation of 5.00 g of glucose.   [3]

Volume of oxygen gas .... dm^3

Q15 The complete combustion of the gas ethane, C_2H_6, forms carbon dioxide and water:

$$2C_2H_6 + 7O_2 \longrightarrow 4CO_2 + 6H_2O$$

(a) A mixture of 10 dm^3 of C_2H_6 with 60 dm^3 of oxygen undergoes complete combustion at a temperature of 650 K.

Calculate the total volume of gas present at the end of the reaction.   [3]

Volume of gas at 650 K .... dm^3

(b) Cooling the gas mixture formed in part (a) to a temperature of 300 K causes a significant decrease in the volume of gas.

(i) Give **two** reasons the gas volume will decrease.   [2]

....

....

....

Unit 1 Practice questions

(ii) Calculate the volume of gas at a temperature of 300 K. [3]

Volume of gas at 300 K .... dm³

Q16 Hydrogen has two naturally occurring stable isotopes.

The relative isotopic masses of each isotope are given:

Isotope	Relative isotopic mass
Hydrogen-1 (Protium)	1.0078
Hydrogen-2 (Deuterium)	2.0141

(a) State what is meant by the term *relative isotopic mass*. [1]

....

....

(b) A sample of hydrogen gas, H_2, contains 99.988% hydrogen-1 and 1.1570×10^{-2}% hydrogen-2. Calculate the relative molecular mass of hydrogen gas. Give your answer to an appropriate number of significant figures. [3]

Relative molecular mass ....

Q17 A mass spectrometer can be used to measure the masses of species.

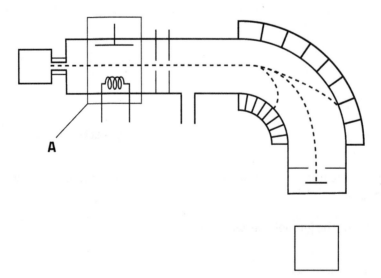

A

(a) Give the name of part A. [1]

....

(b) Explain how the mass spectrometer works. [QER 6]

....

....

....

....

....

....

....

....

....

....

(c) A sample of chloroform, $CHCl_3$, gives four molecular ion peaks at masses of 118, 120, 122 and 124.

(i) Identify the species causing each molecular ion peak. [2]

m/z = 118 ....

m/z = 120 ....

m/z = 122 ....

m/z = 124 ....

(ii) The peak at m/z of 118 has an intensity 27 times greater than that of the peak at m/z of 124.

Explain this observation. [3]

....

....

....

....

Q18 Thionyl chloride, $SOCl_2$ can be used to convert ethanoic acid into ethanoyl chloride:

$$SOCl_2 \quad + \quad CH_3COOH \quad \longrightarrow \quad CH_3COCl \quad + \quad SO_2 \quad + \quad HCl$$
Thionyl chloride + Ethanoic acid $\longrightarrow$ Ethanoyl chloride + Sulfur dioxide + Hydrogen chloride

(a) Calculate the atom economy of this process for preparing ethanoyl chloride. [2]

Atom economy .... %

(b) In an experiment 12.4 g of ethanoic acid forms 11.2 g of ethanoyl chloride. Calculate the percentage yield of this process. [3]

Percentage yield .... %

Q19 The solubilities of magnesium sulfate at different temperatures are given in the table:

Temperature /°C	Solubility in deionised water / mol dm^{-3}
20	2.92
100	4.17

(a) Calculate the mass of anhydrous magnesium sulfate that can dissolve in 250.0 cm^3 of deionised water at 20 °C. [3]

Maximum mass .... g

(b) Cooling 250.0 cm^3 of saturated solution of magnesium sulfate at 100 °C to 20 °C causes crystals of $MgSO_4 \cdot 7H_2O$ to form.

Calculate the maximum mass of crystals that can form. [3]

Maximum mass .... g

(c) Heating the crystals of $MgSO_4 \cdot 7H_2O$ causes them to lose 29.2% of their mass and form a different hydrate, $MgSO_4 \cdot xH_2O$. Find the value of x in this formula. [3]

x = ...

Q20 The maximum mass of anhydrous barium chloride that can dissolve in 100 cm^3 of deionised water is 35.9 g.

(a) State the concentration of this solution in g dm^{-3}. [1]

.... g dm^{-3}

(b) Calculate the concentration of this solution in mol dm^{-3}. [1]

.... mol dm^{-3}

(c) Addition of excess sulfuric acid forms a precipitate of $BaSO_4$.

Calculate the mass of barium sulfate that would be precipitated from 100 cm³ of the barium chloride solution. [2]

Mass of $BaSO_4$ precipitate = ... g

Q21 A student was given a solid unknown monobasic acid, HX, and asked to find its relative molecular mass.

The student dissolved 8.00 g of the solid in 250.0 cm³ of deionised water. He titrated 25.0 cm³ samples against aqueous sodium hydroxide of concentration 0.180 mol dm⁻³ and found 28.15 cm³ of aqueous sodium hydroxide was needed for complete reaction.

Find the relative molecular mass of acid HX. [5]

M_r = ...

Q22 (a) One reaction that occurs in a catalytic converter removes NO and CO.

$$2NO\ (g) + 2CO\ (g) \rightarrow 2CO_2\ (g) + N_2\ (g)$$

In an experiment 10 dm³ of NO was mixed with 15 dm³ of CO and passed through a catalytic converter. Calculate the change in total volume during the experiment. [3]

Change in volume = ... dm³

(b) Trace amounts of sulfur are present in many automotive fuels and when these burn in car engines they release the polluting gas sulfur dioxide, SO_2. To measure the sulfur dioxide in air you can oxidise the sulfur dioxide to sulfate ions, SO_4^{2-} and then precipitate these as barium sulfate.

Apparatus is set up to bubble waste gases from combustion through an oxidising solution. The gas flows through the solution at a rate of 10 dm³ per minute at a temperature of 340 K and a pressure of 1 atm. After 24 hours the solution is treated with barium chloride, and the barium sulfate precipitate separated and dried.

The mass of the precipitate is 0.384 g. Calculate the percentage of SO_2 present in the waste gases. [5]

Percentage sulfur dioxide = ... %

Q23 Zinc metal can displace copper from aqueous copper sulfate:

$$Zn\ (s) + CuSO_4\ (aq) \rightarrow ZnSO_4\ (aq) + Cu\ (s)$$

(a) Copper sulfate exists as the hydrate $CuSO_4.5H_2O$.

(i) Calculate the relative formula mass of this species. [1]

....

(ii) The experiment requires $50.0\ cm^3$ of aqueous copper sulfate of concentration $0.200\ mol\ dm^{-3}$. Calculate the mass of hydrated copper sulfate needed to make this solution. [2]

.... g

(b) A mass of 0.850 g of zinc is added to the $50.0\ cm^3$ of aqueous copper sulfate.

(i) Find which reactant is present in excess. [2]

....

....

(ii) Calculate the maximum mass of copper that can be produced in this reaction. [3]

....

Q24 Heating metal carbonates causes many of them to decompose to form the oxides, releasing carbon dioxide gas. A student conducts an experiment where they heat a known mass of a metal carbonate to 750 K and measure its mass every minute for 5 minutes.

Mass of empty tube	= 39.420 g
Mass of tube + metal carbonate	= 42.262 g
Mass of tube + contents after heating for 1 min	= 41.418 g
Mass of tube + contents after heating for 2 min	= 41.260 g
Mass of tube + contents after heating for 3 min	= 41.249 g
Mass of tube + contents after heating for 4 min	= 41.249 g
Mass of tube + contents after heating for 5 min	= 41.249 g

(a) (i) Calculate the mass of metal carbonate used. [1]

.... g

(ii) Calculate the percentage error in the mass of metal carbonate. [1]

.... %

(iii) State, giving a reason, whether the percentage error in the mass of metal oxide will be greater, less or the same as the percentage error in the mass of metal carbonate. [1]

....

....

(b) The equations for the decomposition of metal carbonates containing M^+ and M^{2+} ions are given below:

$$M_2CO_3 \text{ (s)} \rightarrow M_2O \text{ (s)} + CO_2 \text{ (g)} \qquad\qquad MCO_3 \text{ (s)} \rightarrow MO \text{ (s)} + CO_2 \text{ (g)}$$

Find the relative molecular mass of the metal carbonate and hence identify the metal present. [4]

M_r ....

Metal = ...

Question and mock answer analysis

Q&A 1 Carbon monoxide is a colourless, poisonous gas. One method of measuring the amount of carbon monoxide present in a gas sample is to use its reaction with iodine(V) oxide, I_2O_5:

$$5CO\ (g) + I_2O_5\ (s) \rightarrow 5CO_2\ (g) + I_2\ (s)$$

(a) State, giving a reason, how the volume of gas present will change during this reaction. [2]

(b) The minimum toxic level of carbon monoxide is 100 ppm, that is 100 cm^3 in 1,000,000 cm^3. Calculate the number of moles of carbon monoxide at the minimum toxic level present in 1.0 dm^3 of gas at 298 K and 1 atm pressure. [2]

(c) Calculate the mass of iodine solid that would be formed by reacting 1000 dm^3 of gas at 298 K and 1 atm pressure containing carbon monoxide at the minimum toxic level. [3]

(d) In a separate experiment using a different amount of carbon monoxide, the mass of solid iodine produced is measured on a 3-figure balance. The results recorded are shown:

Mass of empty weighing bottle	= 27.424 g
Mass of weighing bottle containing $I_2(s)$	= 27.661 g

Find the percentage error in the mass of iodine measured. *You must show your working.* [2]

Interpretation

Many calculation questions require you to use the answer for one part in the next, so if you can't get the correct answer to one part, it is important you get an answer, even if you think it is wrong. This will then allow you to move on and use the value in later parts of the question where you can gain full marks for errors carried forward.

Dewi's answer

(a) No volume change (✓) as the number of moles of gas does not change (✓), and the volume of a gas depends on the number of moles only at a constant temperature and pressure.

(b) Vol $(CO) = 100 \div 10^6 \times 1\ dm^3 = 0.0001\ dm^3$ ✓

Moles $= 0.0001 \div 24.5 = 4.08 \times 10^{-6}\ mol$ ✓

(c) Moles $CO = 4.08 \times 10^{-6} \times 1000 = 4.08 \times 10^{-3}\ mol$ ✓

Moles $I_2 = 4.08 \times 10^{-3} \div 5 = 8.16 \times 10^{-4}\ mol$ ✓

Mass $I_2 = 8.16 \times 10^{-4} \times 254 = 0.207g$ ✓

(d) Mass of iodine $= 0.237g$

Error in each measurement $= 0.0005g$ so error in mass $= 2 \times 0.0005g = 0.001g$ ✓

% Error $= 0.001 \div 0.237 \times 100 = 0.4\%$ ✓

MARKER COMMENTARY

(a) This is the correct and complete answer with more detail than needed.

(b) The molar volume is used correctly here.

(c) The calculation uses the value found in part (b) correctly.

(d) The method shows that the error is based on two measurements and the percentage error is calculated correctly. It is often poor practice to give an answer to one significant figure, but in error calculations this is not an issue.

Overall this answer gains 9 marks out of 9 as the calculations are correct and the methods used are clear.

Rebecca's answer

(a) There are the same number of moles (X) on both sides of the equation so there will be no change in volume (✓).

(b) $pV = nRT$ so in $1\ dm^3$ $n = pV \div RT$
$= 1.01 \times 10^5 \div 8.31 \times 298 = 40.8\ mol$ X

As the CO is 100 in 1,000,000 this will be $4.08 \times 10^{-3}\ mol\ CO$ ✓

(c) Moles CO in $1000\ dm^3 = 4.08 \times 10^{-3} \times 1000 = 4.08$ ✓

Moles of $I_2 = 4.08 \div 5 = 0.816\ mol$ ✓

Mass of $I_2 = 0.816 \times 127 = 103.632\ g$ X

(d) Mass of iodine $= 0.237g$

Error in mass = smallest division $= 0.001g$ X

Percentage error $= 0.42\%$ ✓

MARKER COMMENTARY

(a) This is the correct idea and gains a mark, but without the word 'gas' it can't gain the second mark.

(b) Using the ideal gas equation is a harder route to the answer. The method is right but the answer is incorrect as the volume is used in dm^3 rather than the correct m^3

(c) The first two steps use answer (b) correctly and gain 2 ECF marks. The final stage uses the A_r of an iodine atom so does not gain the final mark.

(d) The percentage error is correct but the reason for the error of 0.001 g is not clear so the answer can't gain full marks.

Overall Rebecca gains 5 out of 9 marks as she lacks detail in her explanations and her methods, and she does not make sure her units are correct.

Q&A 2   The relative atomic masses of elements can be found using a mass spectrometer.

(a) State what is meant by the term relative atomic mass. [2]

(b) State and explain the functions of the following parts of the mass spectrometer:

   (i) Electron gun. [2]

   (ii) Negatively charged plates. [1]

(c) The relative atomic mass of rhenium is 186.3 and the mass spectrum shows it shows two isotopes only, with 62.6% being ^{187}Re. Find the mass number of the other isotope. [3]

Interpretation

Most of this question focuses on AO1 (knowledge and understanding). The number of marks guide your answers – in this case part (b)(ii) is worth fewer marks than (b)(i) and (iii) so requires less detail. When studying calculation types, remember that you are expected to understand the method in each case, and not just follow steps you have learned. The calculation of A_r values from given data is a common question, but in this case you need to rearrange the method which makes it more challenging.

Dewi's answer

(a) The relative atomic mass is the mass of one atom of the element (X) compared to 1/12th the mass of an atom of carbon-12. ✓

(b) (i) The electron gun fires high energy electrons at the sample (✓). This knocks off electrons to form positive ions. ✓

(ii) These attract the positive ions and accelerate them. ✓

(c) $A_r = [(\text{abundance} \times \text{mass of isotope 1}) + (\text{abundance} \times \text{mass of isotope 2})] \div 100$

$100 \times A_r - \text{abundance} \times \text{mass of isotope 1} = \text{abundance} \times \text{mass of isotope 2}$ ✓

$(100 \times 186.3) - (62.6 \times 187) = 37.4 \times \text{mass of isotope 2}$ ✓

$6923.8 = 37.4 \times \text{mass of isotope 2}$

mass of isotope 2 = 185 ✓

MARKER COMMENTARY

(a) The use of 1/12th the mass of an atom of carbon-12 is correct for all definitions of relative mass. The idea of relative atomic mass being an average is missing.

(b) (i) The answer includes what the electron gun does and why it does this so gains both marks.

(ii) This question is only worth 1 mark, so the answer is enough to gain this mark.

(c) The answer is clear and complete, leading to the correct answer. The value calculated is 185.1; however, mass numbers are integers so has been correctly stated as 185.

Dewi's answer gains 7 out of 8 marks. He is able to fully describe the way the mass spectrometer functions and perform calculations using data associated with the mass spectrometer.

Rebecca's answer

(a) The relative atomic mass is the average mass of an atom of the element (✓) compared to the mass of carbon-12. ✗

(b) (i) The electron gun fires electrons at high speeds (✓) at the sample to form ions. ✗

(ii) These attract the positive ions and have a small hole that focuses the ions into a beam. ✓

(c) $A_r = \dfrac{(\%\text{age of 1} \times \text{mass 1}) + (\%\text{age of 2} \times \text{mas 2})}{100}$ ✓

$196.3 = \dfrac{(62.6 \times 187) + (32.4 \times \text{mass})}{100}$ ✓

$6923.8 = 37.4 \times \text{mass}$

Mass of isotope = 185.13 ✓

MARKER COMMENTARY

(a) The idea of average mass is key for the relative atomic mass; however, the 1/12th is missing.

(b) (i) The answer includes what the electron gun does but the reference to forming ions is not clear enough.

(ii) This question is only worth 1 mark, so the answer is enough to gain this mark. Reference to acceleration or focusing are both acceptable in this case.

(c) The calculation is correct, and the numerical answer is right. Although mass numbers are better stated as whole numbers, this is not penalised.

Rebecca's answer gains 6 out of 8 marks. Her calculation is correct and she understands how a mass spectrometer works but her answers lack the detail needed for a higher mark.

1.4: Bonding

Topic summary

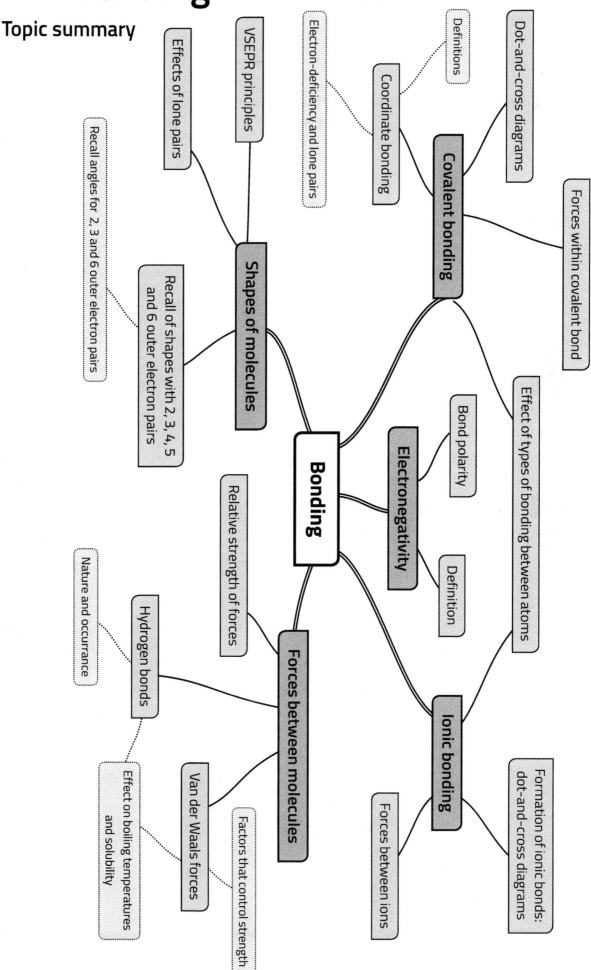

Q1 Place the following bonding types in order of increasing strength. [1]

Covalent bond   Van der Waals forces   Hydrogen bonding

Weakest .... *Strongest*

Q2 Place the following species in order of increasing bond angle. [1]

BCl_3 $BeCl_2$ CCl_4 SF_6 NCl_3

Smallest angle .... *Largest angle*

Q3 Draw a dot-and-cross diagram to show the formation of sodium oxide from atoms of the elements. [2]

Q4 Place the following compounds in order of increasing boiling temperature. [1]

NH_3 F_2 Cl_2 $NaCl$

Lowest .... *Highest*

Q5 The electronegativity values of a selection of elements are given in the table:

Element	H	C	O	F
Electronegativity value	2.1	2.5	3.5	4.0

(a) Give the meaning of the term *electronegativity*. [1]

....

....

(b) Use the electronegativity values to show any dipoles on the bonds below. [2]

O–H   C=O   H–F   C–F

(c) State, giving a reason, which of the bonds in part (b) will have the largest dipole. [2]

....

....

Q6 Three substances including fluorine are hydrogen fluoride, magnesium fluoride and fluorine gas.

(a) Draw a dot-and-cross diagram of F_2. [1]

(b) Sketch the electron distributions in F_2 and hydrogen fluoride and explain the difference. [3]

Fluorine, F_2 .... *Hydrogen fluoride, HF*

....

....

....

(c) Draw dot-and-cross diagrams to show the formation of magnesium fluoride from atoms of the elements. [2]

Q7 (a) Draw a dot-and-cross diagram to show the bonding in a molecule of phosphine, PH_3. [1]

(b) State the shape of a molecule of PH_3. [1]

....

Q8 Explain why the boiling temperatures of the halogens F_2, Cl_2, Br_2 and I_2 increase down the group. [2]

....

....

....

Q9 Place the following compounds in order of increasing boiling temperature, explaining your reason(s).

[4]

$$CH_3CH_2OH \qquad CH_3CH_2CH_2OH \qquad H_3C-O-CH_3$$
$$A \qquad\qquad B \qquad\qquad C$$

Lowest .... *Highest*

....

....

....

....

Q10 Sodium fluoride is an ionic compound.

(a) Use a dot-and-cross diagram to show the bonding present in sodium fluoride. [2]

(b) Sketch the electron density distribution in sodium fluoride. [1]

(c) Explain why ionic compounds such as sodium fluoride have high melting temperatures. [2]

....

....

....

....

Q11 Hydrogen gas, H_2, and helium gas, He, are the elements with the lowest boiling temperatures.

(a) Draw a dot-and-cross diagram for the hydrogen molecule, H_2. [1]

(b) Describe the forces present in a covalent bond. [3]

....

....

....

....

(c) There are Van der Waals forces between molecules of hydrogen. Explain how these forces form and why they lead to such a low boiling temperature for hydrogen. [3]

....

....

....

....

....

Q12 Sulfur forms a range of compounds with fluorine including sulfur difluoride, SF_2, and sulfur hexafluoride, SF_6.

Use valence shell electron pair repulsion theory (VSEPR) to find and draw the shapes of these two molecules. Explain your reasoning. [5]

....

....

....

....

....

Q13 Ammonia, NH_3, forms hydrogen bonding and this affects its physical properties.

Explain fully what is meant by hydrogen bonding and how it affects the properties of ammonia. [QER 6]

....

....

....

....

....

....

....

....

....

....

....

....

Q14 Boron trifluoride, BF_3, is an electron-deficient species.

(a) Draw a dot-and-cross diagram of boron trifluoride and use it to explain the meaning of the term electron-deficient. [2]

....

....

....

(b) Boron trifluoride can combine with ammonia by forming a coordinate bond.

State what is meant by a coordinate bond. [2]

....

....

....

(c) (i) Give the F–B–F bond angle present in BF_3. .... [1]

(ii) Explain why the F–B–F bond angle decreases when BF_3 forms a coordinate bond. [1]

....

....

....

(iii) The H–N–H bond angle in ammonia is smaller than the H–C–H bond angle in CH_4. Explain this difference. [2]

....

....

....

Q15 Carboxylic acids contain the functional group –COOH, which contains a C=O double bond and a C–O–H group.

Carboxylic acid	Boiling Temperature / °C
Methanoic acid, H–COOH	101
Ethanoic acid, CH_3–COOH	118
Propanoic acid, CH_3CH_2–COOH	141
Butanoic acid, $CH_3CH_2CH_2$–COOH	
Pentanoic acid, $CH_3CH_2CH_2CH_2$–COOH	186
Hexanoic acid, $CH_3CH_2CH_2CH_2CH_2$–COOH	205

(a) Describe and explain the pattern in the boiling temperatures. [4]

....

....

....

....

....

(b) Suggest a boiling temperature for butanoic acid. .... °C [1]

(c) Place propanoic acid, butanoic acid and pentanoic acid in order of increasing solubility. Explain your answer. [4]

Least soluble .... Most soluble

....

....

....

....

Q16 The boiling temperatures of the hydrides of some group 6 elements are given in the table.

Hydride	Boiling temperature /°C
H_2O	100
H_2S	−60
H_2Se	−41
H_2Te	−2

Describe and explain the pattern in these boiling temperatures. [5]

....

....

....

....

....

Question and mock answer analysis

Q&A 1 Aluminium forms a range of compounds including $AlCl_3$ and Al_2O_3.

(a) Aluminium chloride is a covalent compound whilst aluminium oxide is ionic.
Explain the difference in bonding between these compounds. [2]

(b) Aluminium chloride can exist as monomers, $AlCl_3$, or dimers, Al_2Cl_6.

    (i) Draw a dot-and-cross diagram of $AlCl_3$. [1]

    (ii) Give the shape of the $AlCl_3$ molecule. [1]

    (iii) Explain why the $AlCl_3$ molecule can form dimers (Al_2Cl_6). [2]

Interpretation

When answering AS questions it is important to remember that some topics are based on the same ideas as GCSE whilst others require a different approach. In this case a GCSE-style answer to parts (a) and (b)(iii) will not gain marks, whilst for (b)(i) a GCSE answer would be awarded the mark.

Dewi's answer

(a) Ionic bonding occurs where two elements which bond together have a large difference in electronegativity, and where the difference is smaller, the bonding is covalent (✓). The difference in electronegativity between Al and O is greater than Al and Cl so aluminium oxide is ionic but aluminium chloride is covalent (✓).

(b) (i)

✓

(ii) Triangular planar. ✓

(iii) The two $AlCl_3$ molecules join together by coordinate bonds (✓). This is because the $AlCl_3$ molecule has lone pairs which can join to the other electron-deficient $AlCl_3$ molecule (✗).

MARKER COMMENTARY

(a) Difference in electronegativity is the key idea in explaining bonding types. It is applied correctly to the two molecules in the question so gains both marks.

(b) (i) The diagram is correct. This is a GCSE-style representation which remains acceptable at AS and gains the mark.

(ii) The diagram in part (i) shows three electron pairs and this is the correct shape for a three electron-pair molecule.

(iii) The type of bonding is correctly identified. Although the answer references the key points of lone pairs and electron deficiency it does not link these with specific atoms so does not gain the second mark.

Dewi's answer is awarded 5 out of 6 marks. He is able to apply the ideas of electronegativity and bonding correctly to the molecules; however, a lack of precision in the final part loses him a mark.

Rebecca's answer

(a) Aluminium is a metal and oxygen is a non-metal so they form ionic bonding (X). Aluminium is near the zig-zag line between metals and non-metals so it can form covalent bonding in some compounds (X).

(b) (i)

Cl   X

Cl • Al
      x
    • x
   Cl

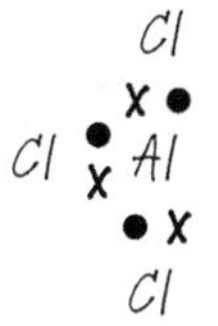

(ii) Trigonal planar. ✓

(iii) Aluminium is electron deficient because it has fewer than 8 outer electrons. Chlorine has lone pairs (✓). These lone pairs join to the aluminium to form coordinate bonds to join two aluminium chloride molecules together (✓).

MARKER COMMENTARY

(a) This is a GCSE-style answer referencing metals and non-metals and is not of the standard expected at AS so gains no marks.

(b) (i) The diagram lacks the lone pairs on the chlorine atoms so does not gain the mark.

(ii) The diagram in part (i) shows three electron pairs around aluminium so the shape predicted is correct.

(iii) This is a good answer, with the aluminium identified as electron deficient and the lone pairs on the chlorines being noted, even though they are not in the diagram in part (i). The type of bonding is correctly identified and explained.

Rebecca's answer is awarded 3 out of 6 marks. In work that overlaps with GCSE she does not make the step up needed for AS, but in the new work on shapes and coordinate bonding she gains marks.

Q&A 2 Boron trichloride, BCl_3, and nitrogen trichloride, NCl_3 have different shapes. Use valence shell electron pair repulsion (VSEPR) theory to find the shapes of these two molecules, explaining the differences between them. [6 QER]

Interpretation

When discussing VSEPR it is important to discuss the principles of the theory and the numbers of both bonding pairs and lone pairs, even when lone pairs are not present. Another common error in many comparison answers is to focus exclusively on what is different and to ignore those things that are unchanged. This is a Quality of Extended Response (QER) question, so marks are awarded on the overall standard of the answer and the way it is expressed.

Dewi's answer

Valence Shell Electron Pair Repulsion theory is based on the idea that electron pairs in the outer shells of atoms will repel each other until they are as far from each other as possible, to minimise the repulsion forces (✓). Boron trichloride has six electrons on the central boron atom, so it has three bonding pairs and no lone pairs (✓). This gives it a trigonal planar shape, with angles of 120° (✓). A nitrogen atom has 5 electrons, and with the 3 covalent bonds it has 8 electrons in the outer shell. These form 3 bond pairs and one lone pair (✓). Electron pairs repel each other, and lone pairs repel more than bonded pairs (✓). This gives the molecule a pyramidal shape (✓) with angles that are less than the usual tetrahedral angle of 109.5°.

MARKER COMMENTARY
This answer includes the principles of VSEPR and this includes the additional repulsion of lone pairs. The numbers of bond pairs and lone pairs in both molecules are correct and this gives the correct shape for each molecule.
The answer is awarded 6 out of 6 marks. The content is complete and is expressed clearly and correctly so it can gain the top band of marks (5–6 marks).

Rebecca's answer

To use VSEPR we must count the number of electron pairs in an atom, and these will space themselves to minimise repulsion. Bonded pairs repel equally, but pairs not involved in bonding (called lone pairs) repel more (✓). Each molecule has three lone pairs, but nitrogen trichloride has one lone pair (✓). This gives a trigonal planar shape for BCl_3 (✓) and a distorted tetrahedral structure for NCl_3 (✗).

MARKER COMMENTARY
This answer includes some of the principles of VSEPR; however, it neglects the idea of outer electrons, but this includes the additional repulsion of lone pairs. This is sufficient for one mark. The numbers of bond pairs in both molecules are correct as is the one lone pair for NCl_3. It is not clear how many lone pairs are present in BCl_3. Once again this is a statement with much correct but parts unclear and it contributes one mark. The correct shape is given for BCl_3 but the description of NCl_3 is incorrect giving one mark for this part.
The answer is awarded 3 out of 6 marks. It is expressed appropriately in most parts so with the standard of the content it would be a mid-band answer (3–4 marks).

1.5: Solid structures

Topic summary

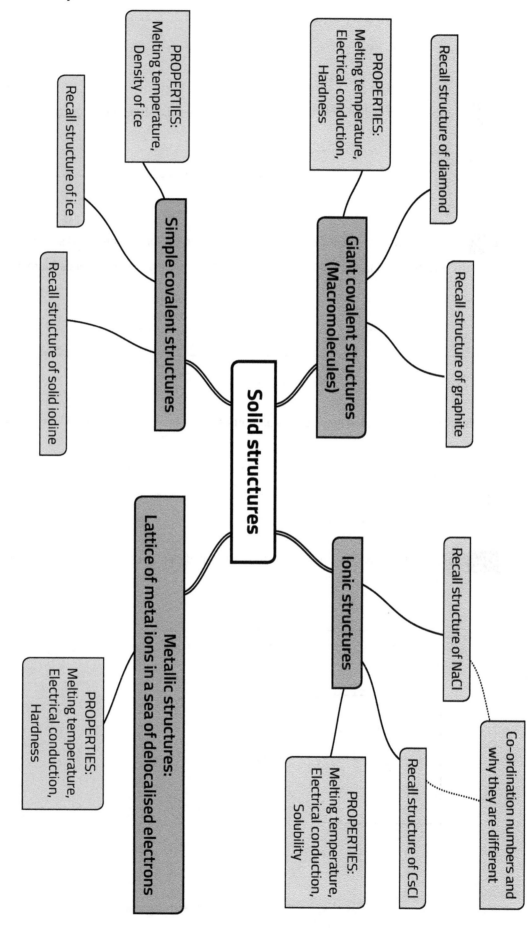

Q1 Diamond and iodine are both elements than contain covalent bonds.

Explain why the boiling temperature of iodine is much lower than the boiling temperature of diamond.

[2]

....

....

....

Q2 Explain why sodium chloride is an insulator in the solid state but conducts electricity when molten. [2]

....

....

....

Q3 Both caesium metal and caesium chloride contain Cs^+ ions in their solid structures.

(a) Give one similarity and one difference between the solid structures of caesium metal and caesium chloride.

[2]

....

....

....

(b) Give one difference between the properties of caesium metal and caesium chloride. [1]

....

....

....

Q4 (a) Complete the diagram below to form a labelled diagram showing the arrangement of ions in sodium chloride.

[2]

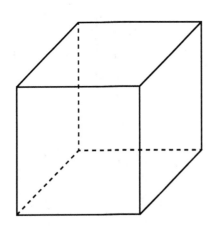

(b) Sodium chloride and calcium oxide both have the same arrangement of ions.

(i) State, giving a reason, whether calcium chloride could have the same arrangement of ions. [1]

....

....

....

....

(ii) The melting temperature of calcium oxide is 2572 °C whilst sodium chloride has a melting temperature of 801 °C. Explain why the melting temperature of calcium oxide is much higher than sodium chloride. [2]

....

....

....

....

(c) Sodium chloride is soluble in water.

Draw a diagram to show the interactions between the ions and water molecules. [2]

Q5 Diamond and graphite are both forms of the element carbon.

(a) Briefly describe the similarities and differences between the bonding and structures of diamond and graphite. [4]

....

....

....

....

....

....

(b) Explain why both graphite and diamond have very high melting temperatures. [2]

....

....

....

(c) Explain why graphite can conduct electricity. [1]

....

....

Q6 Iodine and ice both form crystals containing small molecules.

(a) Tick the boxes to indicate whether each type of bonding is present crystals of these two substances. [2]

	Ionic bonds	Covalent bonds	Coordinate bonds	Van der Waals forces	Hydrogen bonds
Iodine, I_2 (s)					
Water, H_2O (s)					

(b) Explain why iodine has a relatively low boiling point. [2]

....

....

....

(c) When most substances freeze, their density increases; however, when water freezes to form ice, its density decreases.

(i) Suggest why the density of most substances increases when they freeze. [2]

....

....

....

(ii) Explain why the density of water decreases when it freezes to form ice. [3]

....

....

....

....

Q7 Use the ideas you have studied in this unit to explain the following statements.

(a) Group 1 metals such as sodium and potassium are soft whilst those of group 2 such as magnesium and calcium are much harder. [3]

....

....

....

....

....

(b) The best conductors of electricity include metals which are at the end of the *d*-block such as copper, silver and gold. [2]

....

....

....

....

Question and mock answer analysis

Q&A 1 Graphite, aluminium and caesium chloride can all conduct electricity under the correct conditions. Describe briefly the bonding in each of these, explaining how this allows the substance to conduct electricity.

You should state clearly any conditions required for the substance to conduct electricity. **[6 QER]**

Interpretation

When a question requires a brief description there is no advantage in producing a long and detailed discussion – you may find that this reduces the time available to answer other questions. In this case the description should focus on the parts of the bonding that allow electrical conductivity.

Dewi's answer

Graphite is a covalent macromolecule with layers of carbon atoms, each bonded covalently to three others at an angle of 120° to form hexagons. The layers are held together by weak Van der Waals forces. Between the layers are delocalised electrons – these are the fourth valence electrons of each carbon, with the three others in covalent bonds. It is the delocalised electrons that allow graphite to conduct electricity. When a potential difference is applied, the electrons can move and act as a current. This conduction happens in the solid state. ✓✓

Aluminium has metallic bonding. The aluminium forms ions (Al^{3+}) releasing three electrons each. The ions form an ordered lattice structure. The released electrons form a sea of delocalised electrons which allow all metals to conduct electricity. When a potential difference is applied the electrons move and a current flows. This conduction happens in the solid and liquid states. ✓✓

Caesium chloride is an ionic substance. It contains Cs^+ and Cl^- ions and these are arranged in a body centred cubic structure. The ions are held together by strong electrostatic forces between oppositely charged ions. There are no delocalised electrons and in the solid state the ions are held in fixed positions so they cannot move and conduct electricity. When the solid is melted, or when it is dissolved, the ions become free to move and they can then carry charge and the substance becomes a conductor. ✓✓

MARKER COMMENTARY
This answer is complete and detailed; however, it includes much more information that the 'brief description' required and gains no additional credit for this. The answer is set out appropriately and clearly. This answer therefore gains 6 marks out of 6 as it covers all the required content in a clear and well-structured format.

Rebecca's answer

Graphite has layers of carbon atoms bonded by covalent bonds and forming a hexagonal pattern. Between the layers are delocalised electrons. Aluminium is a metal, so it has a lattice of positively charged metal ions in a sea of delocalised electrons. Caesium chloride has an ionic lattice with the Cs^+ and Cl^- ions held in an ordered pattern. ✓✓✓

Both graphite and aluminium have delocalised electrons so can conduct as these electrons can move when a voltage is applied (✓). Caesium chloride only conducts when in the molten state as this frees up the electrons to move and carry a current. (✗)

MARKER COMMENTARY
This answer is structured very differently from Dewi's answer and is much more concise. The descriptions of the structures in each case are brief but sufficient for this question.

The references to delocalised electrons in graphite and aluminium are correct; however, there is no reference to the fact that these two can conduct in the solid state. The statement that caesium chloride only conducts when molten is acceptable, but the reason refers to delocalised electrons rather than ions moving. This is a common error and is penalised in the marking.

This answer therefore gains 4 marks out of 6 as it is structured appropriately and covers the structures well but contains errors and omissions in the discussion of conduction.

Q&A 2 (a) (i) Draw a labelled diagram of the structure of caesium chloride. [2]

(ii) Sodium chloride and caesium chloride have different coordination numbers. Give the coordination numbers of both and explain why they are different. [2]

(b) Graphite, iodine and water all contain covalent bonds. Place these three substances in order of **increasing** boiling temperature, explaining your answer. [3]

Interpretation

These questions require you to make comparisons between different substances and use these to put substances in order. When making comparisons you must refer to all the substances, and use comparisons throughout – greater, smaller, higher, lower rather than large, small, high or low.

Dewi's answer

(a) (i)

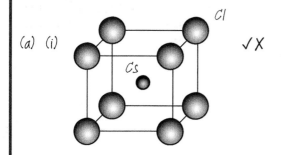

✓ X

(ii) NaCl has 6:6 coordination numbers and CsCl has 8:8 (✓). These are larger in caesium chloride as Cs⁺ is larger than Na⁺ and so more Cl⁻ can fit around it. ✓

(b) Iodine < water < graphite ✓

Iodine and water contain covalent bonds but we only need to break the intermolecular forces to turn them into gases. For water these include hydrogen bonds but for iodine there are only the weaker Van der Waals forces of intermolecular attraction (✓). This gives iodine a lower boiling point than water. To boil graphite you need to break covalent bonds (✓), and these are much stronger than the intermolecular forces in water and iodine, giving graphite a much higher boiling temperature.

MARKER COMMENTARY

(a) (i) The diagram gains only one mark as it is not correctly labelled – the species present are ions so should be Cs⁺ and Cl⁻ rather than Cs and Cl.

(ii) The coordination numbers are identified correctly, and the reason given is clear and correct.

(b) This answer gains all three marks. The order of the boiling temperatures is correct, and the comparison of water and iodine builds on the ideas in the unit of work on bonding. The reason for the high boiling temperature of graphite identifies the need to break covalent bonds which is often overlooked.

Dewi gains 6 out of 7 marks. His explanations are complete, and he makes clear comparisons between the different substances.

Rebecca's answer

(a) (i)

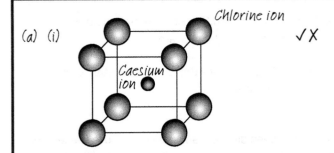

Chlorine ion

✓ X

Caesium ion

(ii) The coordination number of sodium chloride is 6 and caesium chloride is 8 (X). The coordination number for Cs^+ is larger because the Cs^+ is a larger ion than Na^+ and so more chlorine ions can fit around each one. ✓

(b) Iodine, water , graphite ✓

The molecules in graphite are the layers and these are very big so the Van der Waals forces between them are much stronger than usual. This gives graphite a much higher boiling point than the other two which contain small molecules (X).

Comparing water and iodine: In the case of water the intermolecular forces include hydrogen bonds but for iodine there are only the weaker Van der Waals forces of intermolecular attraction. This gives iodine a lower boiling point than water. ✓

MARKER COMMENTARY

(a) (i) The diagram gains only one mark as it is not correctly labelled: the anion is a chloride not chlorine.

(ii) The coordination numbers are given for just one ion, so no mark. The reason given is clear and as the use of chlorine ion has already been penalised it is accepted here.

(b) This answer gains two marks. The order of the boiling temperatures is correct, and the comparison of water and iodine correctly discusses the intermolecular forces and their relative strength. The reason given for the higher boiling temperature of graphite is incorrect.

Rebecca gains 4 out of 7 marks. Most of her explanations are correct; however, she misses some details.

1.6: The periodic table

Topic summary

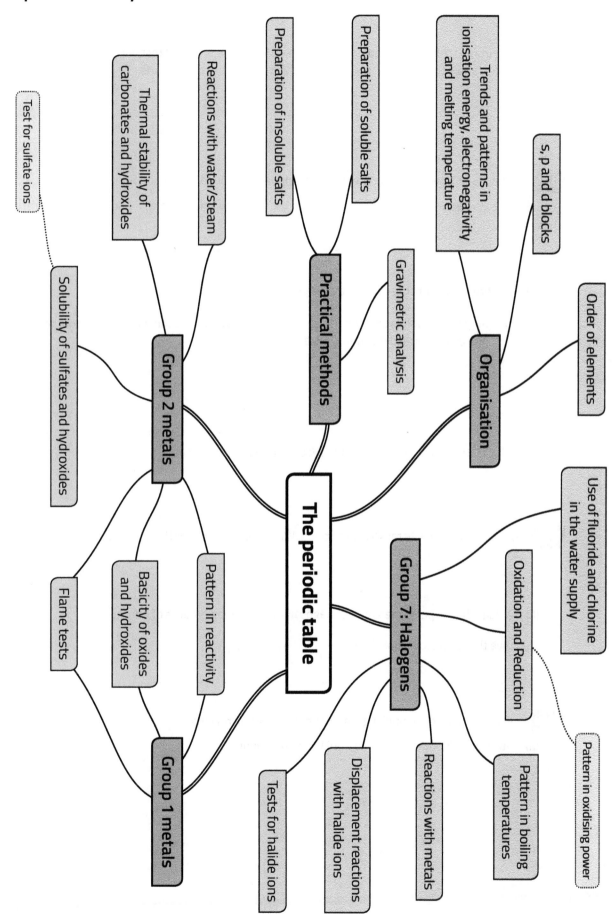

Q1 Iron reacts with chlorine gas to form iron(III) chloride.

(a) Write the equation for this reaction. [1]

(b) The iron(III) chloride formed is dissolved in water. Give a chemical test, including the expected observation, that would show that the solution contains chloride ions. [2]

Q2 Two students wished to test whether a solution contained barium ions.

(a) One student performed a flame test. Give the expected colour, if any, if barium ions are present. [1]

(b) The other student chose to use a precipitation reaction.

(i) Suggest a suitable solution to use in this precipitation reaction, giving the colour of the precipitate formed if barium ions are present. [2]

(ii) Write an ionic equation for the reaction occurring in part (i). [1]

Q3 (a) Magnesium reacts rapidly with steam. Write the equation for this reaction. [1]

(b) A flame test is performed on the magnesium-containing product of part (a).

Give the expected colour, if any. [1]

Q4 Fluoride is added to drinking water supplies in some regions.

(a) State why fluoride is added to drinking water supplies. [1]

(b) The salt sodium fluorosilicate, Na_2SiF_6, is one compound used to fluoridate water supplies. All the fluorine atoms are released as fluoride in the process.

A water supplier wishes to increase the fluoride content of water from 7.0×10^{-6} mol dm^{-3} to 3.9×10^{-5} mol dm^{-3}. Calculate the mass of sodium fluorosilicate that will need to be added to each dm^3. [4]

Mass of Na_2SiF_6 = ... g

Q5 The third period of the periodic table consists of the elements shown:

Group	1	2	3	4	5	6	7	0
Element	Na	Mg	Al	Si	P	S	Cl	Ar

(a) Identify, giving a reason, the element with the highest melting temperature. [1]

....

....

(b) Identify, giving a reason, the element with the highest first ionisation energy. [1]

....

....

....

(c) Identify, giving a reason, the element with the highest electronegativity. [1]

....

....

Q6 Over 1.6 million tonnes of chlorine, Cl_2, are produced annually in the UK and it has a wide range of uses.

(a) Calculate the number of moles of chlorine produced annually in the UK. [2]

.... mol Cl_2

(b) Chlorine is classed as a *p*-block element. State what is meant by a *p*-block element. [1]

....

....

(c) State why chlorine in used in the production of drinking water. [1]

....

....

(d) Chlorine can be used to produce bromine from the bromide ions in seawater.

(i) Write the ionic equation for this reaction. [1]

....

(ii) Explain why chlorine can oxidise bromide ions, whilst iodine cannot. [2]

....

....

....

(e) Chloride is present in a range of minerals. One such mineral is carnallite $KMgCl_3.xH_2O$.

(i) State what would be seen if carnallite was used in a flame test. [1]

....

(ii) Heating a sample of carnallite causes it to lose water and become anhydrous. During the process carnallite loses 38.9% of its mass. Calculate the value of x in the formula $KMgCl_3.xH_2O$. [3]

X = ...

(iii) A sample of carnallite was dissolved in water. State what would be observed, if anything, when the following solutions are added to a solution of carnallite. [2]

Solution added	Observation
Aqueous sodium hydroxide	
Dilute sulfuric acid	
Aqueous silver nitrate	

Q7 A student is provided with nine bottles labelled A–I containing the following aqueous solutions:

Sodium nitrate, Sodium hydroxide, Sodium carbonate,

Potassium sulfate, Potassium bromide, Potassium iodide,

Potassium nitrate, Barium chloride, Magnesium nitrate

In addition, the student has bottles of the following reagents:
- Dilute nitric acid
- Aqueous silver nitrate
- Concentrated ammonia solution.

She also has the apparatus required for performing flame tests.

Plan a method of identifying all nine solutions conclusively, giving the expected observations at each stage. You should aim to produce a plan that is as efficient as possible.

[QER 6]

Q8 (a) Magnesium and barium are two metals in group 2.

State and explain which of these two metals is the more reactive. [2]

(b) Sodium and magnesium are two metals in the same period.

State and explain which of these two metals has the highest melting temperature. [2]

....

....

....

....

(c) The oxides of sodium, magnesium and barium are classed as basic oxides.

State what is meant by the term basic oxide. [1]

....

....

(d) Magnesium carbonate, $MgCO_3$, is heated gradually until it reaches constant mass. At temperatures between 400 °C and 800 °C the mass of the sample decreases to 47.8% of its original mass as carbon dioxide is released.

(i) Write an equation for the reaction occurring. [1]

....

(ii) The experiment is repeated using barium carbonate in place of magnesium carbonate.

(I) Suggest the temperature needed for decomposition of barium carbonate. [1]

....

(II) Calculate the percentage of the original mass remaining after heating a sample of barium carbonate to constant mass. [2]

.... %

(e) A student is asked to prepare a solid sample of the soluble salt barium chloride, starting from aqueous barium hydroxide and dilute hydrochloric acid. The student plans the following method:

- Measure 25 cm³ of the aqueous barium hydroxide in a measuring cylinder.
- Place the solution in a conical flask.
- Add a few drops of a phenolphthalein indicator.
- Fill a burette with the hydrochloric acid.
- Add the hydrochloric acid to the aqueous barium hydroxide 1 cm³ at a time until the indicator changes colour.
- Place the solution in an evaporating basin and heat until the solution is saturated.
- Cool the solution and filter away the crystals.
- Dry the crystals in an oven until their mass is constant.

The teacher says that the method needs to be changed. Suggest the changes needed to the method, explaining why these are needed. [5]

....

....

....

....

....

....

....

....

Q9 This question focuses on the elements listed in the section of the periodic table below:

1	2										3	4	5	6	7	0
													N		F	
Na	Mg											Si	P		Cl	
K	Ca														Br	
	Sr														I	
Cs	Ba															

You can only use the elements labelled above. You may use each element once, more than once, or not at all.

Identify:

(a) An element with a half-filled *p* sub-shell. .... [1]

(b) An element that has a crimson flame test. .... [1]

(c) An element that forms a stable ion with a −1 charge. .... [1]

(d) The metal with the highest melting temperature. .... [1]

(e) The element with the smallest electronegativity value. .... [1]

(f) The least reactive *s*-block metal. .... [1]

(g) The halogen that is the strongest oxidising agent. .... [1]

Q10 A solution is known to contain a mixture of potassium carbonate and potassium nitrate only.

(a) Give a test which would confirm the presence of potassium ions in this solution. State the result of the test. [2]

....

....

....

(b) To determine the carbonate ion concentration in the mixture, aqueous calcium chloride was added in excess. The precipitate produced was filtered off and dried by heating.

(i) Write the ionic equation, including state symbols, for this precipitation reaction. [1]

....

(ii) State why the aqueous calcium chloride was added in excess. [1]

....

....

(iii) The M_r of hydrated calcium carbonate produced in this experiment was 208.2. When 10.00 g of the sample was heated the mass decreased in stages as the temperature increased. At different points in the heating the mass of the sample was 5.67 g and 4.81 g until it ends up as 2.69 g at the end of the heating.

Explain the changes occurring during the heating, using calculations to confirm the species present at each stage. [5]

....

....

....

....

....

(iv) A teacher stated that using barium chloride to form a precipitate of barium carbonate would allow the student to get more accurate results.

Explain why this is the case. [2]

....

....

....

....

Question and mock answer analysis

Q&A 1 This question focuses on the elements listed in the section of the periodic table below.

1	2									3	4	5	6	7	0
Li															
Na	Mg									Al	Si	P		Cl	
K	Ca					Fe		Cu						Br	
														I	
Cs	Ba														

You can only use the elements labelled above. You may use each element once, more than once, or not at all.

Identify:

(a) A *d*-block element.

(b) An element in the third period.

(c) An element that forms a stable ion with a +2 charge.

(d) An element that forms a basic oxide.

(e) The non-metal with the highest melting temperature.

(f) The element with the greatest electronegativity value.

(g) The most reactive *s*-block metal.

(h) A metal with an insoluble carbonate and a soluble sulfate.

(i) The group 2 metal with the most thermally stable carbonate.

Interpretation

Questions such as these are almost all AO1 marks. They test recall of facts and applying these to simple contexts. These questions either start 'The' or 'A' and it is important to realise that there is only one correct answer to questions starting 'The' but more than one possible answer for those starting 'A'. In each case you should only give one element as the answer, as if you give two answers and one is incorrect and one correct you do not get a mark.

Dewi's answer

(a) Fe ✓

(b) Na ✓

(c) Fe ✓

(d) Na ✓

(e) Si ✓

(f) Cl ✓

(g) Cs ✓

(h) Ba ✗

(i) Ba ✓

MARKER COMMENTARY
The first seven elements chosen are all correct; however, the eighth is incorrect. Barium forms an insoluble carbonate; however, its sulfate is also insoluble. The stability of the carbonates increases down the group so the answer to part (i) is correct. Due to excellent recall of the factual content Dewi gains 8 out of 9 marks.

Rebecca's answer

(a) Cu ✓

(b) K ✗

(c) Mg ✓

(d) Mg ✓

(e) Fe ✗

(f) Cl ✓

(g) K ✗

(h) Cu ✓

(i) Mg ✗

MARKER COMMENTARY

Three of the first four elements chosen are all correct; however, the element potassium is in the fourth period not the third. This is a common misconception as the first period, containing hydrogen and helium is missing from the section of the periodic table shown.

The selection of iron for part (e) shows that the question has not been read correctly as it asks for a non-metal. In part (g) learners are familiar with potassium as the most reactive of the s-block metals they commonly encounter; however, the reactivity increases down the group so caesium will be even more reactive. The other answers are all correct answers.

Rebecca gains 5 out of 9 marks as she recalls a significant amount of factual content but lacks care in its application. She is less able to apply trends to make predictions for less familiar elements.

Q&A 2 Norsethite is a mineral of formula $A_aB_b(CO_3)_c$ where A and B are s-block metals. A student performed a series of tests on the mineral to find the formula of the mineral:

	Test	Result
1	Addition of excess nitric acid to 1.90×10^{-3} mol of the hydrated mineral to form a solution.	The sample fizzes producing a volume of 85.1 cm^3 of gas at 273 K and 1 atm pressure.
2	Addition of excess sodium hydroxide to the solution formed in test 1.	A white precipitate forms. Separating the precipitate and heating this to constant mass gives 76.6 mg of a metal oxide.
3	Flame test	An apple green flame is seen

(a) Write the **ionic** equation for the reaction of carbonate ions with acid. [1]

(b) State what information is obtained from the flame test. [1]

(c) Calculate the number of moles of carbon dioxide gas produced in test 1 and hence calculate the value of c, number of carbonate ions in the formula. [3]

(d) (i) Suggest which metal ions are present in the precipitate in test 2. [1]

(ii) Find the number of these metal ions present and hence give the formula of norsethite. [4]

Interpretation

Questions such as these are mainly AO3 marks as they require you to combine information and come to a conclusion – in this case the formula of norsethite. If you make an error, or can't answer a part, carry on as the formula will be marked consequentially.

Dewi's answer

(a) CO_3^{2-} (aq) + $2H^+$(aq) $\longrightarrow$ CO_2 (g) + H_2O (l) ✓

(b) The sample contains barium ions (✓) and does not contain any other ions that give flame colours.

(c) Number of moles = $85.1 \times 10^{-3} \div 22.4$ ✓

   = 3.80×10^{-3} mol ✓

   c = $3.80 \times 10^{-3} \div 1.90 \times 10^{-3}$ = 2 ✓

(d) (i) The ion could be magnesium (✓) or calcium.

   (ii) If there is one metal ion in the mineral then the M_r of the oxide will be
   $76.6 \times 10^{-3} \div 1.90 \times 10^{-3}$ = 40.3.
   This would be the M_r of MgO. ✓✓

   As there are two carbonate ions these have 4– so there must be 1 Ba and 1 Mg to give 4+. The formula of norsethite is $MgBa(CO_3)_2$ ✓✓

MARKER COMMENTARY

(a) The equation is correct.

(b) Barium is correct and would be enough as an answer.

(c) Dewi correctly changes units and uses these correctly to find the value of c.

(d) (i) Although magnesium is the expected answer, calcium is also accepted.

   (ii) The calculation is one correct way to show the formula contains one Mg ion.
   The formula follows from all the previous information, and the ion charges are balanced so gains both marks.

Dewi's answer gains 10 out of 10 marks. His answers include all the correct information, and he is able to combine all this information to gain the formula of norsethite.

Rebecca's answer

(a) CO_3^{2-} + $2 HNO_3$ $\longrightarrow$ CO_2 (g) + H_2O + $2 NO_3^-$ ✗

(b) The sample contains barium ions. ✓

(c) $pV = nRT$ so $n = pV/RT = 1 \times 85.1 / 8.31 \times 273$

   = 3.75×10^{-2} mol (✓ ECF)

   c = $3.75 \times 10^{-2} / 1.90 \times 10^{-3}$ = 20 (✓ ECF)

(d) (i) The ion could be magnesium. ✓

   (ii) The M_r of magnesium oxide is 40.3 so 76.6×10^{-3} is 1.90×10^{-3} mol. ✓

   This is the same as the number of moles of compound so there must be 1 Mg in the formula. ✓

   The formula of norsethite is $MgBa(CO_3)_{20}$ ✓ ✗

MARKER COMMENTARY

(a) The equation incorrectly includes HNO_3 rather than H^+.

(b) Barium is correct.

(c) Rebecca uses the ideal gas equation correctly; however, the units of pressure and volume are incorrect. The moles calculated follow from this error, as does the value of c so she gains these marks.

(d) (i) Magnesium is the expected answer and gains the mark.

   (ii) The calculation is one correct way to show the formula contains one Mg ion.
   The ions present in the formula follow from all the previous information, as does the 1 Mg^{2+} and 20 CO_3^{2-} ions. The charges are not balanced as 19 Ba^{2+} would be needed so only one mark is gained.

Rebecca's answer gains 7 out of 10 marks as she has most of the correct ideas, and where she makes calculation errors, she has used the values correctly in later steps.

1.7: Simple equilibria and acid–base reactions

Topic summary

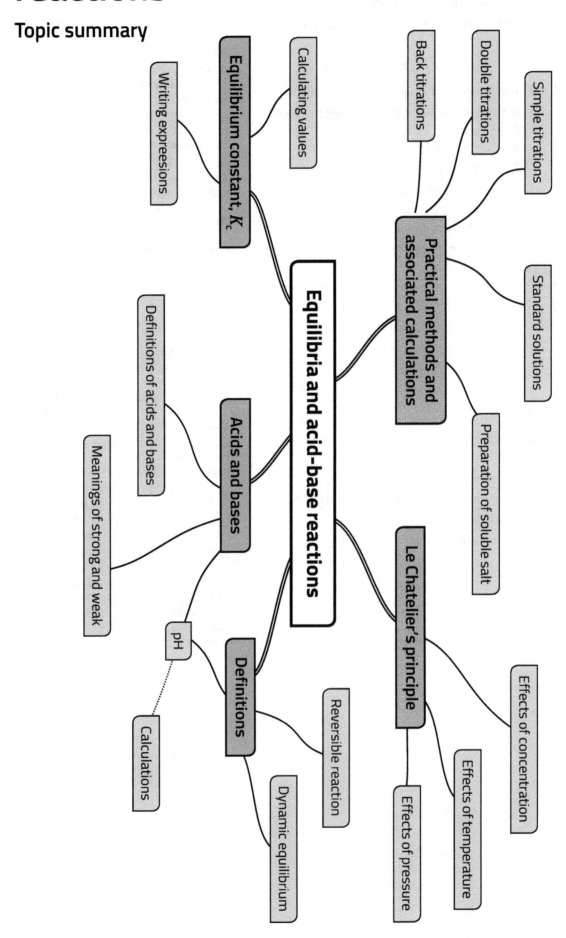

Q1 Define pH. [1]

....

Q2 State what is meant by the term dynamic equilibrium. [1]

....

....

Q3 Write the equilibrium constant, K_c, for the reversible reaction given. [1]

$$2SO_2 + O_2 \rightleftharpoons 2SO_3$$

Q4 Find the pH of a solution of the strong acid HCl of concentration 0.20 mol dm^{-3}. [1]

pH = ...

Q5 An acid with a pH of 3.0 could be a weak acid or a dilute acid. Explain the difference between these terms. [2]

....

....

....

Q6 The weak acid ethanoic acid can be produced by the carbonylation of methanol:

$$CH_3OH + CO \rightleftharpoons CH_3COOH$$

(a) Write the equilibrium constant, K_c, for this reversible reaction. [1]

(b) The pH of a sample of ethanoic acid is 4.35. Find the concentration of H$^+$ ions in this solution. [2]

Concentration of H$^+$ ions = ...mol dm^{-3}

Q7 Hydroiodic acid is a strong acid. It is formed by dissolving hydrogen iodide in water.

(a) A sample of hydroiodic acid is formed by dissolving 15.0 g of HI gas in 250 cm³ of deionised water.

Calculate the pH of this solution. [3]

pH = ...

(b) Hydrogen iodide can be formed from hydrogen and iodine gases in a reversible reaction:

$$H_2 (g) + I_2 (g) \rightleftharpoons 2HI (g)$$

(i) Write the expression for the equilibrium constant in terms of concentration, K_c, for this reversible reaction. [1]

....

(ii) Under certain conditions the concentrations of hydrogen and iodine are both 0.65 mol dm⁻³ and the equilibrium constant is 45.0. Calculate the concentration of hydrogen iodide. [3]

Concentration of HI = ...mol dm⁻³

(iii) The reaction is repeated at a higher temperature, and the equilibrium constant is found to be 46.5. State and explain what information this provides about the energy change of the reaction. [3]

....

....

....

(iv) The pressure of the reaction vessel is increased. State, giving a reason, the effect of this on the yield of hydrogen iodide. [2]

....

....

....

Q8 In solution in $CHCl_3$, N_2O_4 forms NO_2 in a reversible reaction:

$$N_2O_4 \rightleftharpoons 2NO_2$$

(a) Write an expression for the equilibrium constant, K_c, for this reaction. [1]

(b) 2.53 g of N_2O_4 was dissolved in 100 cm³ of $CHCl_3$ and the mixture allowed to reach equilibrium. The mass N_2O_4 in the solution at equilibrium was 0.38 g.

(i) Find the mass of NO_2 in solution and hence calculate its concentration. [3]

Mass of NO_2 = ...g

Concentration of NO_2 = ...mol dm⁻³

(ii) Calculate the concentration of N_2O_4 in the solution and hence calculate the value of K_c under these conditions. *Units are not required.* [2]

Concentration of N_2O_4 = ...mol dm⁻³

K_c = ...(units not needed)

(c) The equilibrium between N_2O_4 and NO_2 can also occur in the gas phase.

State, giving a reason, the effect of increasing pressure on the amount of NO_2 present in the gas mixture at equilibrium. [2]

....

....

....

Q9 Ammonia is produced from hydrogen and nitrogen gases in the Haber process:

$$N_2 (g) + 3H_2 (g) \rightleftharpoons 2NH_3 (g) \qquad\qquad \Delta H = -92 \text{ kJ mol}^{-1}$$

(a) Describe and explain how temperature and pressure will affect the yield of ammonia in this process.

[4]

....

....

....

....

....

(b) Ammonia is extremely soluble in water. Reaction of ammonia solution with sulfuric acid produces soluble ammonium sulfate.

(i) Write an equation for this reaction, including the state symbols. [1]

....

(ii) Outline a method to prepare pure, dry crystals of ammonium sulfate using this reaction. [5]

....

....

....

....

....

....

Q10 The concentration of an acid can be found by titration of the acid against a standard solution of a base, such as sodium carbonate.

(a) A student needs to prepare 250 cm³ of standard aqueous sodium carbonate of concentration 0.100 mol dm⁻³.

(i) Calculate the mass of sodium carbonate required to make this standard solution. [2]

Mass of sodium carbonate = ...mol dm⁻³

(ii) Outline how to prepare the standard solution.

You should name the apparatus required for your method. [4]

....

....

....

....

....

....

....

(b) A student places a 25.0 cm³ sample of the standard sodium carbonate solution in a conical flask and titrates it using hydrochloric acid. A volume of 17.10 cm³ of acid was needed to exactly neutralise the sodium carbonate:

$$Na_2CO_3 + 2HCl \rightarrow 2NaCl + CO_2 + H_2O$$

Calculate the concentration of the hydrochloric acid. [3]

Concentration of HCl = ...mol dm⁻³

Q11 Solid F is a mixture of sodium hydroxide and sodium hydrogencarbonate. This is dissolved in water to form solution G. To find the concentrations of the two substances in solution G a student performed a double titration. They used the following method:

- Accurately measure 25.0 cm³ of solution B and place in a conical flask.
- Add a few drops of phenolphthalein indicator and record the initial reading on the burette. This is called reading A.
- Add the hydrochloric acid whilst swirling the solution. Continue until the solution changes from pink to colourless. At this point all the sodium hydroxide has reacted.
- Record the reading on the burette. This is called reading B.
- Add a few drops of methyl orange and add the hydrochloric acid with swirling.
- Continue until the solution changes colour.
- Record the reading on the burette. This is called reading C. At this point all the sodium hydrogencarbonate has reacted.
- The concentration of hydrochloric acid used was 0.220 mol dm⁻³.

The results of the experiment are given below.

	1	2	3	4
Reading A / cm³	0.20	0.45	0.00	1.40
Reading B / cm³	16.25	15.85	15.55	16.85
Reading C / cm³	24.10	23.70	23.45	24.70
Volume that reacted with NaOH / cm³				
Volume that reacted with NaHCO₃ / cm³				

(a) Complete the table and calculate the mean volumes required for reaction with NaOH and reaction with $NaHCO_3$. [2]

Mean volume that reacts with NaOH = ... cm^3

Mean volume that reacts with $NaHCO_3$ = ... cm^3

(b) (i) One student states that the data show that the value for the volume that reacts with NaOH is more reliable than the value for the volume that reacts with $NaHCO_3$. Give a reason for this. [1]

....

....

(ii) Another student states that the data show that the value for the volume that reacts with $NaHCO_3$ is more reliable than the value for the volume that reacts with NaOH. Give a reason for this. [1]

....

....

(c) (i) Find the concentrations of NaOH and $NaHCO_3$ in solution G. [3]

Concentration of NaOH = ... $mol\ dm^{-3}$

Concentration of $NaHCO_3$ = ... $mol\ dm^{-3}$

(ii) Calculate the percentage of sodium hydrogencarbonate by mass in solid F. [3]

Percentage $NaHCO_3$ = ... %

Q12 Alcohols react with carboxylic acids to form esters in an endothermic reaction.

An example is the formation of the ester ethyl ethanoate:

ethanoic acid + ethanol $\rightleftharpoons$ ethyl ethanoate + water

$$CH_3COOH + CH_3CH_2OH \rightleftharpoons CH_3COOCH_2CH_3 + H_2O$$

(a) State and explain the effect of increasing temperature on the yield of ethyl ethanoate. [2]

....

....

....

(b) Concentrated sulfuric acid is usually added to the mixture. This strong acid acts as a catalyst and a dehydrating agent.

(i) State what is meant by the term strong in this context. [1]

....

....

(ii) Explain why a dehydrating agent will increase the yield of ethyl ethanoate. [2]

....

....

....

(c) Ethanoic acid is a weak acid, so it dissociates, releasing H^+ ions in a reversible reaction:

$$CH_3COOH \rightleftharpoons CH_3COO^- + H^+$$

State and explain the effect of adding concentrated sulfuric acid on the degree of dissociation of ethanoic acid. [2]

....

....

....

Q13 A typical eggshell has a mass of approximately 6 g. Most of this is calcium carbonate, $CaCO_3$.

The calcium carbonate content of the eggshell can be analysed by back titration. An eggshell is treated with 200 cm³ of 1.00 mol dm⁻³ hydrochloric acid and the solution made up to 250 cm³ using deionised water. Samples of 25.0 cm³ of the solution are titrated against aqueous sodium hydroxide of concentration 0.500 mol dm⁻³. The results of the experiment are listed below:

Mass of eggshell = 5.74 g

Volume of aqueous sodium hydroxide = 18.20 cm³

(a) A **two** decimal place balance was used. Calculate the maximum percentage error in the weighing of the eggshell.

You **must** show your working. [2]

Maximum percentage error .... %

(b) Write the equation for the reaction of calcium carbonate with hydrochloric acid. [1]

....

(c) Calculate the amount of hydrochloric acid added to the eggshell. [1]

.... mol

(d) Calculate the amount of hydrochloric acid remaining in each 25.0 cm³ sample used in the titration. [2]

.... mol

(e) Calculate the percentage by mass of calcium carbonate in the eggshell. [4]

.... %

Q14 Ethanoic acid, CH_3COOH, is a weak acid.

(a) State what is meant by the term *weak acid*. [2]

....

....

(b) Write the equation for the reaction of ethanoic acid with copper oxide. [1]

....

(c) Vinegar is a dilute solution of ethanoic acid. A sample of white vinegar has a pH of 2.7.

(i) Find the concentration of H^+ ions in this solution. [1]

$[H^+]$ = ...mol dm^{-3}

(ii) A 25.0 cm^3 sample of this vinegar reacts completely with 22.10 cm^3 of aqueous sodium hydroxide of concentration 0.250 mol dm^{-3}.

Find the concentration of this acid in g dm^{-3}. [3]

Concentration of ethanoic acid = ...g dm^{-3}

Unit 1 Practice questions

Question and mock answer analysis

Q&A 1 The Fischer-Tropsch reaction is a process that can be used to make a range of hydrocarbons. One example is the production of ethane:

$$5H_2 \text{ (g)} + 2CO \text{ (n)} \rightleftharpoons C_2H_6 \text{ (g)} + 2H_2O \text{ (g)} \quad \Delta_fH^\ominus = -347 \text{ kJ mol}^{-1}$$

(a) This process establishes a dynamic equilibrium.
State what is meant by the term *dynamic equilibrium*.   [1]

(b) Write an expression for the equilibrium constant, K_c, for this reaction.   [1]

(c) State and explain the effect of increasing temperature on the yield of ethane.   [2]

(d) State and explain the effect of increasing pressure on the yield of ethane.   [2]

Interpretation

This question covers the main concepts regarding equilibria in the AS course. The use of Le Chatelier's principle is examined in parts (c) and (d) and it is important to read the question carefully. These questions require you to discuss the yield of ethane, and answers that do not cannot gain full credit.

Dewi's answer

(a) A dynamic equilibrium is a reversible reaction where the rate of the forward and reverse reactions are equal. ✓

(b) $K_c = \dfrac{[C_2H_6][H_2O]^2}{[H_2]^5 [CO]^2}$ ✓

(c) Increasing the temperature causes the equilibrium to shift to the endothermic direction (✓), which is to the left. This reduces the yield of ethane. ✓

(d) Increasing the pressure causes the equilibrium to shift to the side with fewer molecules (✗), which is to the right. This increases the yield of ethane. ✓

MARKER COMMENTARY

(a) This is the correct answer.

(b) This is the correct expression for K_c.

(c) Dewi uses Le Chatelier's principle to explain the effect of increasing temperature, and then uses it to give the effect of this on the yield of ethane. He therefore gains both marks.

(d) Dewi uses Le Chatelier's principle to explain the effect of increasing pressure but refers only to the number of molecules and not to the number of GAS molecules so he doesn't gain the first mark. The effect of pressure on the yield of ethane is correct and gains this second mark. He therefore gains one mark.

Dewi gains 5 out of 6 marks as he can apply Le Chatelier's principle to this equilibrium, as well as treating it quantitatively using an equilibrium constant.

Rebecca's answer

(a) A dynamic equilibrium is a reversible reaction where the forward and reverse reactions are equal. ✗

(b) $K_c = \dfrac{(C_2H_6) + (H_2O)^2}{(H_2)^5 (CO)^2}$ ✗

(c) Increasing the temperature causes the equilibrium to shift to the left as this is the endothermic direction. ✓ ✗

(d) Increasing the pressure causes the equilibrium to shift the right as this is the side with fewer gas molecules. ✓ ✗

MARKER COMMENTARY

(a) This is not sufficient for the mark as there is no reference to rate.

(b) Rebecca has the correct substances in the correct positions but uses round brackets rather than square brackets. It is only square brackets that represent concentration. The values should all be multiplied, so the plus symbol is another error. This does not gain a mark.

(c) She uses Le Chatelier's principle to explain the effect of increasing temperature on the position of equilibrium. This gains the first mark. There is no reference to yield of ethane, so she does not gain the second mark.

(d) Rebecca uses Le Chatelier's principle to explain the effect of increasing pressure on the position of equilibrium. This gains the first mark. There is no reference to yield of ethane, so she does not gain the second mark.

Rebecca gains 2 out of 6 marks as she can apply Le Chatelier's principle but lacks the details needed to gain a greater mark.

Q&A 2 Acids can be classed as strong acids such as nitric acid, HNO_3, or weak acids such as ethanoic acid, CH_3COOH. These acids can be compared using their pH values.

(a) State what is meant by the term *weak acid*. [2]

(b) Define pH. [1]

(c) Find the pH value of nitric acid of concentration 0.200 mol dm^{-3}. [1]

(d) A solution of pH 2.87 could be a weak acid or a very dilute solution of a strong acid. Titration of 50.0 cm^3 of this acidic solution against 5.00×10^{-3} mol dm^{-3} aqueous sodium hydroxide gave the results below:

	1	2	3	4
Volume of NaOH (aq) / cm^3	13.80	13.40	13.45	13.40

(i) Calculate the mean volume of aqueous sodium hydroxide required for complete reaction. [1]

(ii) Calculate the concentration of the acid solution, and hence find whether it is a strong or weak acid. [4]

Interpretation

This question requires an understanding of the differences between strong and weak acids as well as the ideas of concentrated and dilute. It is important that the answers make the link between the calculated values and the definitions of these ideas.

Dewi's answer

(a) A weak acid is one that is partially dissociated in aqueous solution. ✓ ✗

(b) pH = $-\log[H^+]$ ✓

(c) pH = $-\log(0.200) = 0.70$ ✓

(d) (i) 13.4166667 ✓

(ii) moles of NaOH = 13.4166667 × 5.00 × 10^{-3} ÷ 1000
= 6.71 × 10^{-5} mol ✓

Concentration of acid
= 6.71 × 10^{-5} ÷ (50.0 × 10^{-3})
= 0.001342 mol dm^{-3} ✓

Concentration of H$^+$ from pH 2.87
= 10$^{-2.87}$ = 0.001349 mol dm^{-3} ✓

As the concentration of the H$^+$ equals the concentration of the acid, all the acid molecules must be dissociated so it is a strong acid. ✓

MARKER COMMENTARY

(a) Dewi explains what is meant by weak and gains a mark but does not define acid so does not gain the second.

(b) This is the correct answer.

(c) The calculation is correct, as is the answer.

(d) (i) The mean volume calculated is correct as it doesn't include the first value.

(ii) The correct concentration of acid is calculated. The method used to show that the acid is completely dissociated is appropriate and Dewi links this correctly to the strength of the acid.

Dewi gains 8 out of 9 marks as he can recall and apply his knowledge of acids and related calculations.

Rebecca's answer

(a) A weak acid is one that dissociates reversibly in aqueous solution. ✓ X

(b) $pH = -\log[H^+]$ ✓

(c) $pH = -\log(0.200) = 0.70$ ✓

(d) (i) $13.51 \, cm^3$ (X)

(ii) moles of NaOH = $13.51 \times 5.00 \times 10^{-3} \div 1000$
= 6.76×10^{-5} mol (✓ ECF)

Concentration of acid = $6.76 \times 10^{-5} \div 50.0$
= 1.342×10^{-6} mol dm^{-3} X

pH of this solution if it is strong
= $-\log (1.342 \times 10^{-6}) = 5.87$ (✓ ECF)

As the pH is the solution is lower than the pH it would have if fully dissociated, then the acid can't be fully dissociated, and it must be a weak acid. X

MARKER COMMENTARY

(a) The answer explains what is meant by weak but does not refer to H^+ ions.

(b) This is the correct answer.

(c) The calculation is correct, as is the answer.

(d) (i) The answer uses the first value despite it being too far away.

(ii) The moles of NaOH are calculated using the correct method. The concentration of the acid is a factor of 1000 out due to incorrect units.

Her method for finding if the acid is strong is valid but the explanation does not gain the second mark as it implies a weak acid would have a lower pH than a strong acid of the same concentration.

Rebecca's answer gains 5 out of 9 marks. She makes several calculation errors but carries on and gets marks for her method with the errors carried forward.

Unit 2: Energy, Rate and Chemistry of Carbon Compounds

2.1: Thermochemistry

Topic summary

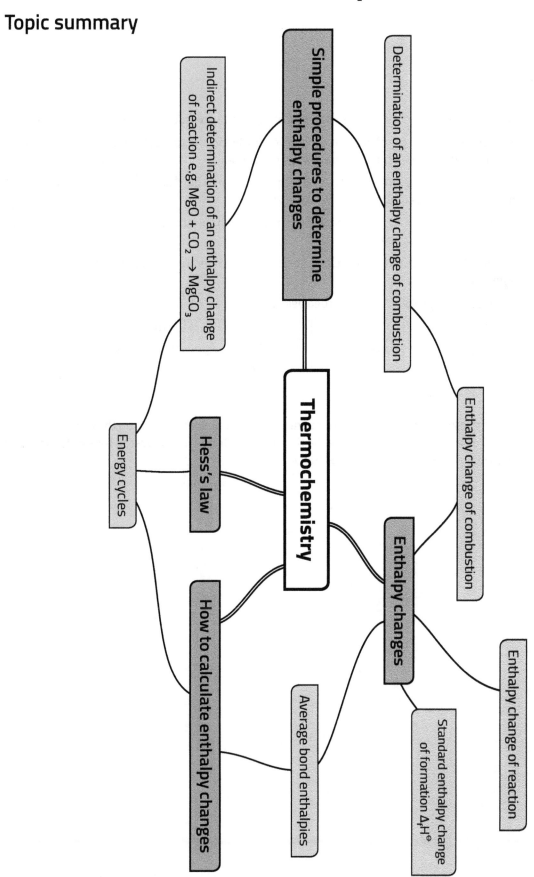

Draw the energy profile for the reaction below. Label the activation energy and the enthalpy change on your diagram.   [2]

$$CH_4(g) + H_2O(g) \rightleftharpoons CO(g) + 3H_2(g) \qquad \Delta H = +206 \text{ kJ mol}^{-1}$$

 A student carried out a simple laboratory experiment to investigate the enthalpy of neutralisation.

They transferred 25.0 cm³ of 1.00 mol dm⁻³ hydrochloric acid to a polystyrene cup. They then stirred the acid and measured the temperature every minute.

In the meantime, they measured out 25.0 cm³ of 1.00 mol dm⁻³ sodium hydroxide solution and measured its temperature – the value was 20.2 °C.

At the third minute they added the sodium hydroxide to the acid and stirred the mixture well but did not take the temperature.

They continued to stir the mixture and measure the temperature every minute up to and including the eighth minute.

The results they collected are recorded below:

Time / minutes	0	1	2	3	4	5	6	7	8
Temperature /°C	20.4	20.3	20.3	X	26.8	26.6	26.4	26.2	26.0

(a) Use the results to plot a graph of temperature against time on the grid on the next page. Draw a line of best fit for the points before the third minute. Draw a line of best fit for the points after the third minute. Extrapolate both lines to the third minute.   [4]

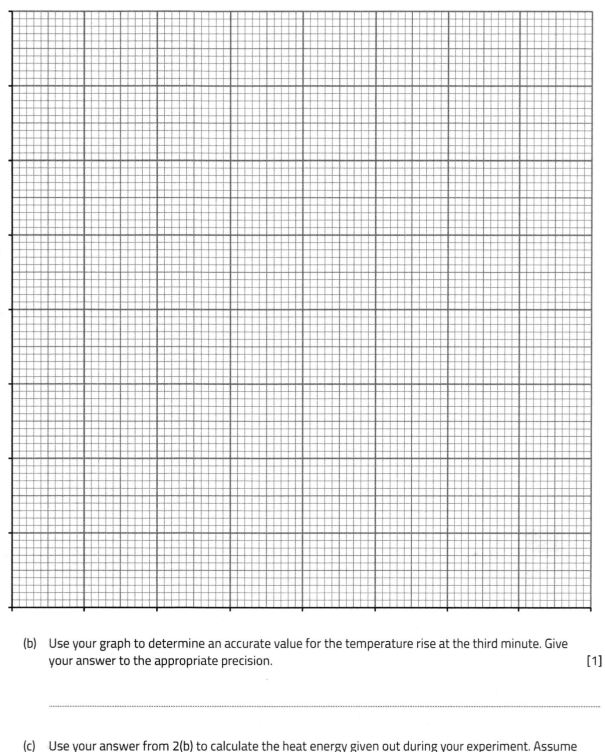

(b) Use your graph to determine an accurate value for the temperature rise at the third minute. Give your answer to the appropriate precision. [1]

....

(c) Use your answer from 2(b) to calculate the heat energy given out during your experiment. Assume that the reaction mixture has a density of 1.00 g cm^{-3} and a specific heat capacity of 4.18 J K^{-1} g^{-1}. Assume that all the heat given out is used to heat the reaction mixture. Show your working. [3]

....

....

....

....

....

....

(d) In the experiment 25.0 cm³ of 1.00 mol dm⁻³ hydrochloric acid were neutralised by 25.0 cm³ of 1.00 mol dm⁻³ sodium hydroxide solution.

(i) Write an equation for this reaction. [1]

....

(ii) Use your answer from 2(c) to calculate a value, in kJ mol⁻¹, for the enthalpy of neutralisation of hydrochloric acid and sodium hydroxide. [3]

....

....

....

....

....

....

(e) Suggest a value for the enthalpy of neutralisation if the hydrochloric acid was exchanged for nitric acid. Explain your answer. [2]

....

....

....

Q3 Carbon undergoes combustion when heated to a high temperature in oxygen, and carbon dioxide can be formed.

(a) State the meaning of the term *standard enthalpy of combustion*. [2]

....

....

....

(b) Under different conditions the product is carbon monoxide. Write an equation, including state symbols, for this reaction. [1]

....

(c) Suggest why it is not possible to measure directly the enthalpy change for this reaction. [1]

....

....

(d) Use the standard enthalpies of formation in the table below and the equation to calculate a value for the standard enthalpy of combustion of pentane. Show your working out. [3]

Species	$\Delta_f H^\circ$ / kJ mol^{-1}
$C_5H_{12}(l)$	-174
$O_2(g)$	0
$CO_2(g)$	-393
$H_2O(l)$	-286

$$C_5H_{12}(l) + 8O_2(g) \rightarrow 5CO_2(g) + 6H_2O(l)$$

....

....

....

....

....

(e) State why the value quoted in part (d) for the standard enthalpy of formation of $CO_2(g)$ is the same as the value for the standard enthalpy of combustion of carbon. [1]

....

....

Q4 The following equation represents the complete combustion of ethanol.

$$CH_3CH_2OH + 3O_2 \rightarrow 2CO_2 + 3H_2O$$

(a) Use the average bond enthalpies from the table below to calculate a value for the standard enthalpy of combustion of ethanol. Show your working out. [4]

Bond	Average bond enthalpy/kJ mol^{-1}
C–H	412
C–C	348
C–O	360
O–H	463
C=O	743
O=O	496

....

....

....

....

....

(b) Give one reason why the value calculated from average bond enthalpies is different from the value given in a data book. [1]

....

....

Question and mock answer analysis

Q&A 1   A student was studying the chemistry of propan-1-ol, which is a liquid at room temperature.

(a) Write an equation for the combustion of propan-1-ol.   [2]

(b) The enthalpy change for this reaction can be determined by creating an energy cycle involving standard enthalpy of formation values.

   Write an equation that represents the standard enthalpy of formation of propan-1-ol.   [2]

(c) Why would it be difficult to measure this value directly?   [1]

(d) Use the standard enthalpy change of formation values given in the table below to calculate the standard enthalpy change for the combustion of propan-1-ol.   [3]

Species	$\Delta_f H^\ominus$ / kJ mol^{-1}
$CO_2(g)$	−394
$H_2O(l)$	−286
$O_2(g)$	0
$CH_3CH_2CH_2OH(l)$	−315

(e) State why the standard enthalpy of formation for oxygen is zero.   [1]

Interpretation

There are several definitions of different enthalpy changes covered in this topic – it is important that you know the wording and are able to convert that wording into an equation (as required here). Understanding these definitions is also important as it helps you answer questions like parts (b) and (d) as well.

Throughout this section there is a requirement to be able to calculate enthalpy changes and this is just one of the common examples – a calculation involving enthalpy of formation values.

Ruairi's answer

(a) $CH_3CH_2CH_2OH + 5O_2 \rightarrow 3CO_2 + 4H_2O$ ✓X

(b) $3C + 4H_2 + 0.5O_2 \rightarrow CH_3CH_2CH\ OH$ ✓X

(c) It's too difficult to measure. X

(d) ΔH = Products − Reactants

   = $(3 \times -394) + (4 \times -286) - -315$

   = $-2011\ kJmol^{-1}$ ✓✓✓

(e) Oxygen is an element. X

MARKER COMMENTARY

(a) Ruairi gave the correct formulae for the reactants and products but did not balance the number of oxygen molecules correctly. He probably overlooked the fact that there is already an oxygen in the formula of the alcohol. 1 mark.

(b) Ruairi correctly gave an equation showing the formation of 1 mole of propan-1-ol from its constituent elements but he did not use state symbols to signify standard states. Only 1 mark.

(c) This answer lacks specific detail. 0 marks.

(d) Ruairi gains full marks for getting the correct value. Strictly speaking, the first line of his work is not correct. He should have referred to the enthalpies of formation rather than just writing 'products', etc., but this can be ignored here because he then used the values correctly. 3 marks.

(e) Ruairi has the right idea but needs to be more precise and he needs to refer to standard states. 0 marks.

Overall Ruairi gains 5 out of 9. Ruairi has good calculation skills which saved him here. He needs to improve his recall of definitions and then use them to help make his answers more precise/chemically correct.

Jessica's answer

(a) $CH_3CH_2CH_2OH + 4.5O_2 \rightarrow 3CO_2 + 4H_2O$ ✓✓

(b) (i) $3C(s) + 4H_2(g) + 0.5O_2(g) \rightarrow CH_3CH_2CH_2OH(l)$ ✓✓

(c) There is no water (or solution present) so we cannot measure the temperature change. ✓

(d) ΔH = Sum of enthalpies of formation of products – Sum of enthalpies of formation of reactants

$= [(3 \times -394) + (4 \times -286)] - (-315)$

$= -2011 \, kJ \, mol^{-1}$ ✓✓✓

(e) Oxygen is an element in its standard state and so there is no enthalpy of formation. ✓

MARKER COMMENTARY

(a) Jessica gave the correct products and correctly balanced the equation. 2 marks.

(b) Jessica gained 2 marks because she gave an equation showing the formation of 1 mole of propan-1-ol from its constituent elements plus she used state symbols to show all the chemicals were in their standard states.

(c) This is a well-worded answer – Jessica has probably come across this question before and has learned a suitable way to answer it. 1 mark.

(d) Jessica lays her answer out well, showing her reasoning, inserting in the values correctly and calculating the answer correctly. 3 marks.

(e) Another precisely worded answer – the essential aspect is that Jessica recognises that the element must be in its standard state, so by definition $\Delta_f H$ is 0. 1 mark.

Overall Jessica gains 9 out of 9 marks. Her good recall of definitions enables her to give all the detail required in these answers. Her calculation skills are also good. An alternative, acceptable approach to these calculations is to use an energy cycle.

Q&A 2 A student was investigating the combustion of hexane.

(a) Give the meaning of the term *standard enthalpy of combustion*. [2]

(b) Use the data in the table below to determine the enthalpy of combustion for hexane. Give your answer in kilojoules. [4]

Mass of water	100 g
Mass of hexane and burner before	64.890 g
Mass of hexane and burner after	65.535 g
Temperature of water before	20.5 °C
Temperature of water after	60.5 °C

(c) The literature value for this enthalpy change is almost double the value that the student obtained.

   (i) What would be the greatest source of error during the experiment? [1]

   (ii) Suggest a way in which this error could be reduced. [1]

(d) Explain why burning the hexane for longer would have given the student a more accurate result. [2]

Interpretation

This is one of the required practicals in this section of the specification. It is essential to understand what is going on in the method as well as being able to describe it. In this question you have to be able to process the data and then recall how to use it to calculate a molar enthalpy change, once you have calculated the heat released.

Ruairi's answer

(a) The enthalpy change when 1 mole of a substance is completely combusted. ✓ ✗

(b) $q = mc\Delta T$

$= 100 \times 4.18 \times 40 = 16.7$ kJ ✓

Mass of hexane $= 65.535 - 64.890 = 0.645$ g

no. of moles $= 0.645/86 = 7.5 \times 10^{-3}$ ✓

$\Delta H = -q/n = -16.7/7.5 \times 10^{-3}$ ✓

$= 2227$ kJ/mol ✓

(c) (i) Heat loss. ✓

(ii) They could put cotton wool around the beaker of water they are heating. ✗

(d) The experiment would have lasted longer and they would have got a larger temperature change which would have helped. ✗ ✗

MARKER COMMENTARY

(a) Ruairi has left off the end phrase 'under standard conditions' so only 1 mark.

(b) Ruairi completes the calculation correctly. Answers like these are usually given within certain ranges. He needs to take care about when and where he rounds his answers because he could end up being 'out of range'. 4 marks.

(c) Correct reason for error but cotton wool would most likely catch fire and is not a sensible improvement. Only 1 mark.

(d) Ruairi's answer is too non-specific. 0 marks

Overall Ruairi gains 6 out of 10 marks. Ruairi recognises and can do these type of calculations but he needs to improve his understanding of this practical and how to calculate errors.

Jessica's answer

(a) The enthalpy change when 1 mole of a substance is burned completely in oxygen under standard conditions. ✓ ✓

(b) $q = mc\Delta T$

$= 100 \times 4.18 \times 40 = 16720$ J ✓

Mass of hexane $= 65.535 - 64.890 = 0.645$ g

number of moles $= 0.645/86.14 = 7.49 \times 10^{-3}$ ✓

$\Delta H = -q/n = -16720/7.49 \times 10^{-3}$ ✓

$= 2232962$ J/mol $= 2233$ kJ/mol ✓

(c) (i) There is a lot of heat lost to the environment. ✓

(ii) They could put a screen around the apparatus to minimise heat lost from the burner flame. ✓

(d) Measurement error is calculated by:

(Uncertainty / Mass reading) $\times$ 100 ✓

So the larger the mass of fuel burned the smaller the error. ✓

MARKER COMMENTARY

(a) Jessica has stated the full meaning correctly. 2 marks

(b) Jessica has completed the calculation well and sensibly leaves rounding and converting units to the end. 4 marks.

(c)(i) and (ii) Correct answer with a sensible and suitable improvement. 2 marks

(d) Jessica knows how errors are calculated and understands how small values lead to large errors. 1 mark.

Overall Jessica gains 10 out of 10. She has a good understanding of this practical and the associated calculations. She also knows how errors are calculated and how results can affect them.

2.2: Rates of reaction

Topic summary

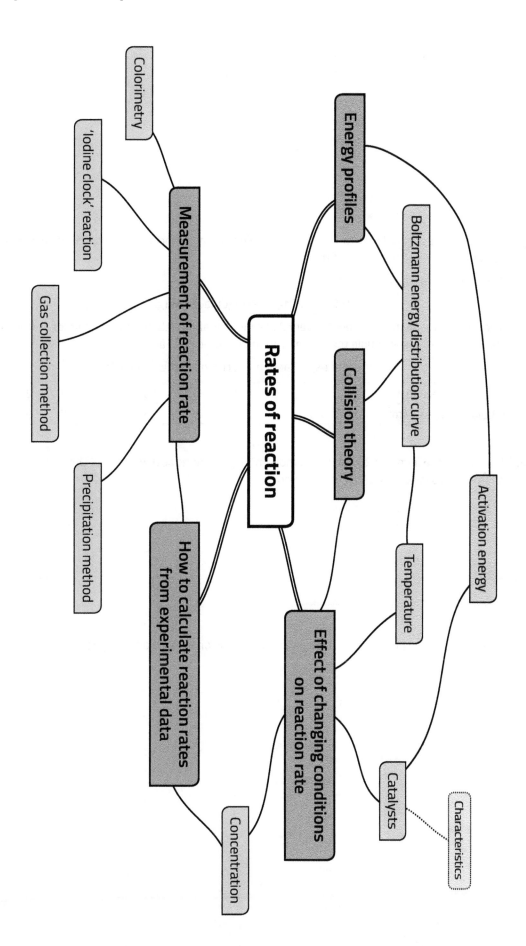

Unit 2 Practice questions

Q1 Hydrogen peroxide decomposes to produce water and oxygen. Sketch and label the apparatus that could be used to measure the rate of this reaction. [2]

Q2 A student investigated how the rate of reaction between hydrochloric acid and magnesium at 20 °C is affected by the concentration of the acid.

The equation for the reaction is:

$$2HCl(aq) + Mg(s) \rightarrow MgCl_2(aq) + H_2(g)$$

The student made measurements every 3 seconds for about 0.5 minutes. The student then repeated the experiment using double the concentration of hydrochloric acid.

The results obtained for the first experiment are recorded in the table below:

Time / seconds	0	3	6	9	12	15	18	21	24	27
Volume of gas / cm³	0	23	38	47	53	55	55	55	55	55

(a) 0.5 g of magnesium and 25 cm³ of 2.0 mol dm⁻³ hydrochloric acid were used. Determine which chemical was in excess. Show your working. [2]

....

....

....

....

(b) Use the results to plot a graph of volume against time on the grid on next page. Draw a line of best fit. [4]

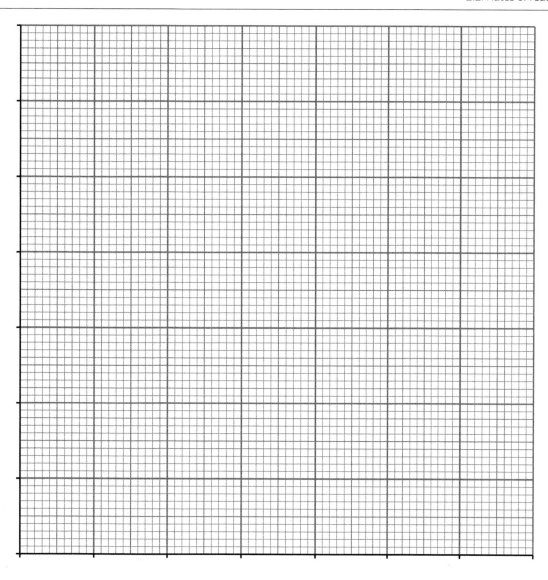

(c) Use your graph to calculate the initial rate. Show all your working. [2]

....

....

....

....

(d) Work out the overall rate for the reaction and explain why this answer is different from the value you calculated in part (c). [3]

....

....

....

....

(e) Explain in detail, using collision theory, why the graph plateaus off after about 12 seconds. [2]

....

....

....

....

Q3 The Boltzmann distribution represents the spread of molecular energies in a gas at a specific temperature. Draw, with labelled axes, a curve to represent this distribution and label this curve Higher T. On the same axes, draw a second curve to represent the same sample of gas at a lower temperature. Label this curve Lower T.

Use these curves to explain why a small decrease in temperature can lead to a large decrease in the rate of a reaction. [8]

....

....

....

....

....

....

....

Q4 (a) Give the meaning of the term *rate of reaction*. [1]

....

....

(b) Explain, using collision theory, why decreasing the concentration of one of the reactants leads to a slower reaction. [2]

....

....

....

(c) A reaction happens between two substances A and B. As the reaction proceeds, the solution turns dark blue due to the formation of C. The reaction occurs according to the following equation:

$$A(aq) + B(aq) \rightarrow C(aq) + D(aq)$$

(i) Suggest how the rate of reaction could be studied. [1]

....

....

(ii) Use the data in the table below to determine the relationship between the concentration of reactant A and the rate of the reaction. Show your working. [2]

Experiment	Initial concentration of A / mol dm^{-3}	Initial concentration of B / mol dm^{-3}	Initial rate / mol dm^{-3} s^{-1}
1	0.12	0.26	6.30×10^{-4}
2	0.36	0.26	1.89×10^{-3}
3	0.72	0.13	1.89×10^{-3}

....

....

....

....

(iii) Does changing the concentration of B affect the rate of reaction? Explain your answer. [2]

....

....

....

....

Q5 Catalysts play an important role in many reactions.

(a) State the meaning of the term *catalyst*. [1]

....

....

(b) Explain, in general terms, how catalysts work. [2]

....

....

(c) Explain what is meant by a homogeneous catalyst. [1]

....

....

(d) Name the catalyst used in the production of esters. [1]

....

(e) Explain why catalysts are frequently used in industrial processes. [1]

....

....

Question and mock answer analysis

Q&A 1 The diagram below shows a Boltzmann distribution curve for the particles in a sample of gas at a given temperature.

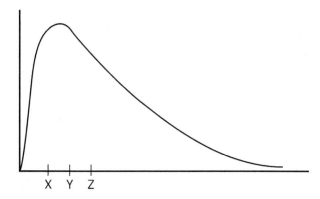

(a) Give the labels for each axis. [2]

(b) Explain why the curve starts at the origin. [1]

(c) State which one of X, Y or Z best represents the most probable energy of the particles. [1]

(d) Explain why, even for a fast reaction, most collisions are not successful. [2]

(e) Describe the shape of the distribution if the reaction was carried out at a higher temperature. [2]

(f) What aspect of the distribution will remain the same if the reaction was carried out at a higher temperature? Explain your answer. [2]

Interpretation

The Boltzmann energy distribution is an important aspect of this topic. You need to learn how to draw it (including labels like most probable energy, where the activation energy is, etc.) and understand what it means. You also need to understand how changing the temperature affects the shape of the distribution. Understanding this distribution will help you explain many questions about how certain conditions affect the rate of reaction.

Ruairi's answer

(a) Energy and number of molecules. ✓X

(b) All of the molecules have some energy. ✓

(c) Y ✓

(d) Most particles don't have enough energy to overcome the activation energy. ✓X

(e) The new curve would be to the right and lower. ✓X

(f) The peak area would stay the same because the molecules are the same. ✓X

MARKER COMMENTARY

(a) Ruairi knows what the axes labels should be but only gains 1 mark due to lack of specificity – which axis is which? Plus he should be careful to avoid using words like 'molecules' if they are not mentioned in the question.

(b) Ruairi correctly explains why the graph starts at zero – his wording is different from Jessica's but still correct. 1 mark.

(c) He correctly picks the letter linked to the highest point of the peak. 1 mark.

(d) Ruairi knows this answer but only gets 1 mark because he does not give sufficient detail about where the activation energy is on the distribution. He could have drawn it on the diagram to be completely clear.

(e) Ruairi has learned this answer but again needs to be clearer in his wording to make it obvious how his peak/curve relates to the original one. Again a quick sketch would have helped – it is okay to include diagrams in such answers. 1 mark for having some correct content.

(f) Ruairi has the right idea but needs to be more precise in the wording of his explanation.

Overall Ruairi gains 6 out of 10 marks. Ruairi has a reasonable understanding of this work but loses marks for lack of precision with his answers. He needs to find ways to express himself clearly and labelled diagrams are useful in doing this. He could also learn from mark schemes – learning the phrasing that they use.

Jessica's answer

(a) x-axis is energy and y-axis is number of
  particles. ✓✓

(b) There are no particles with zero energy. ✓

(c) Y ✓

(d) The activation energy is always quite a lot to
  the right of Y and so most particles have less
  than the activation energy. ✓✓

(e) The peak would be to the right of where Y is
  and lower than the current peak. The curve
  would descend (above the other curve at all
  places) and would not cross the x-axis. ✓✓

(f) The area under the peak would be the same
  because no particles have been added or
  removed. ✓✓

MARKER COMMENTARY

(a) Jessica has labelled the axes correctly and is careful to use the same wording 'particles' as in the question. 2 marks.

(b) Jessica understands the shape of the distribution well. 1 mark.

(c) She correctly picks the letter linked to the highest point of the peak. 1 mark.

(d) Jessica words her answer well and gives sufficient detail for 2 marks. It is important to state where the activation energy is on the distribution as it is not shown in the question.

(e) Jessica words her answer carefully so it is clear how her high temperature peak relates to the other one. 2 marks

(f) Jessica is correct and explains her answer well.

Overall Jessica gains 10 out of 10. She understands these distributions well and is careful when she words her answers to be precise and avoid ambiguous comments.

 Q&A 2 Sodium thiosulfate reacts slowly with hydrochloric acid to form a precipitate. The equation for the reaction is:

$$Na_2S_2O_3 + 2HCl \rightarrow 2NaCl + SO_2 + S + H_2O$$

(a) State which of the products is insoluble and forms a precipitate. [1]

(b) The rate of this reaction can be studied by measuring the time that it takes for a small fixed amount of precipitate to form under different conditions. Describe how this reaction could be carried out to study the effect of temperature on the rate of reaction. [3]

(c) Give the meaning of the term *activation energy*. [1]

(d) Explain why increasing the temperature of the reactants increases the rate of reaction dramatically. [2]

(e) What effect does increasing the temperature of the reaction have upon the activation energy? [1]

Interpretation

This is another of the practical methods that students should be familiar with. The method for recording data to study rates of reactions depends upon the nature of the chemicals involved. In this case, one of the products is a solid and so timing the disappearance of a cross on a piece of paper would work. It would also be possible to use a light meter and measure the drop in light intensity passing through the reaction with respect to time. Make sure you are familiar with all of these methods.

Ruairi's answer

(a) Sulfur is the precipitate. ✓

(b) The reaction is stood on a piece of
paper with a cross on it. The reaction
is timed until the cross disappears.
The reaction is repeated for different
temperatures. ✓✓ X

(c) This is the energy needed for a reaction
to start. X

(d) Increasing the temperature increases
the energy of the particles and so
there will be more successful collisions
per second. ✓ X

(e) It will be easier to overcome. X

MARKER COMMENTARY

(a) Ruairi correctly identifies the solid. 1 mark.

(b) He knows how this reaction should take place and gets some credit for describing the set up (a labelled diagram would help). What he omits from his answer is any comment about controlling the other factors that could affect the rate of reaction. This is an important aspect of such practicals. 2 marks.

(c) Ruairi makes the common mistake of omitting the word 'minimum' from his answer. 0 marks.

(d) Ruairi understands what is going on here correctly, but he should have also referred to the activation energy in his answer, especially so because it was mentioned in part (c). One question often leads onto another so follow the flow. Only 1 mark.

(e) Ruairi misses the point of this question – it's about the value of the activation energy. 0 marks.

Overall Ruairi gains 4 out of 8 marks. Ruairi is aware of this type of practical task and just needs to take more care when describing how to carry it out, and be clear that only one factor is changed and everything else is kept the same. Taking care with wording over key terms is also something he needs to perfect.

Jessica's answer

(a) Sulfur is insoluble. ✓

(b)

view from the top

without precipitate   with precipitate

The reaction and apparatus are set up as shown
above. The reaction is timed until the cross
disappears. The second time this is done the
temperature of sodium thiosulfate is increased (with
all the other quantities/values staying the same) and
the reaction is timed to the same end point – the
same density of precipitate. This can be repeated for
as many temperatures as necessary. ✓✓✓

(c) This is the minimum energy needed for a reaction to
happen. ✓

(d) When the temperature is increased the particles gain
energy and many more particles now have greater
than or equal to the activation energy and so there
will be many more successful collisions per second.
✓✓

(e) No effect. ✓

MARKER COMMENTARY

(a) Jessica correctly identifies the solid formed. 1 mark.

(b) Jessica uses a labelled diagram to help explain how she would carry out this method and is careful to refer to fair testing aspects. 3 marks.

(c) Jessica gives a precise answer including all key words. 1 mark.

(d) Jessica follows on from her answer to part (c) and gives a detailed answer linking increased temperature and activation energy. 2 marks

(e) Only catalysts affect the value of activation energy and Jessica spots this. 1 mark.

Overall Jessica gains 8 out of 8. Good knowledge and understanding of this practical and how it works, alongside well-worded answers enable Jessica to get all the marks.

Unit 2 Practice questions

2.3: The wider impact of chemistry

Topic summary

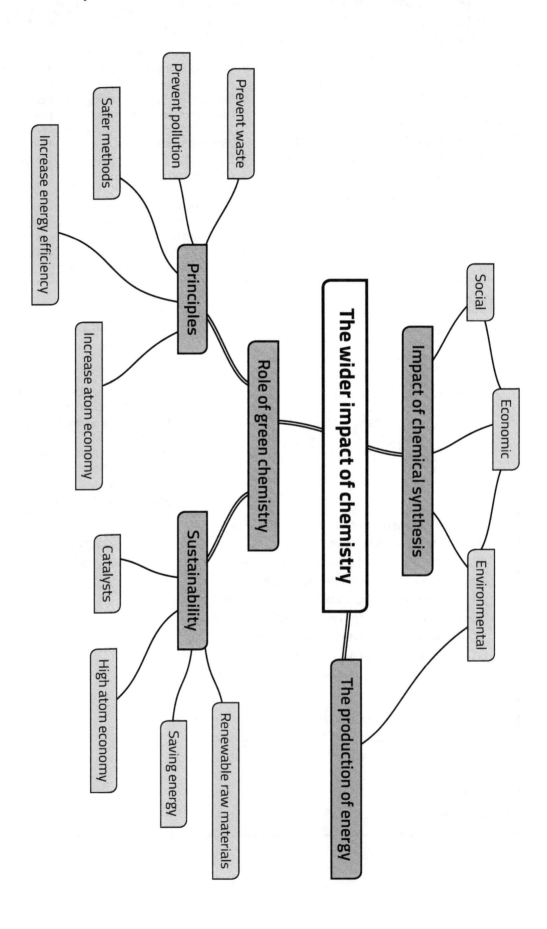

Q1 Ethanol can be produced by fermentation or by hydration.

Fermentation:

$$C_6H_{12}O_6(aq) \rightarrow 2CO_2(g) + 2CH_3CH_2OH(aq)$$

Fermentation is catalysed by enzymes from yeast at a temperature of 40 °C under atmospheric pressure.

Hydration:

$$C_2H_4(g) + H_2O(l) \rightarrow CH_3CH_2OH(aq)$$

Hydration is typically carried out using a catalyst of phosphoric acid at 300 °C and 60–70 atmospheres.

Compare the two processes in terms of the principles of 'green' chemistry. [6]

....

....

....

....

....

....

....

....

....

....

....

Q2 A new chemical plant is being built in your local area. Nearby residents have been asked to contribute suggestions as to what should be considered when constructing this plant – to ensure that it is in 'harmony' with the local environment and community.

What four key points would you include in your response? Explain your ideas. [8]

Question and mock answer analysis

Q&A 1   The reactions below show two ways to produce methanal:

A $CH_3OH(g) \longrightarrow \overset{H}{\underset{H}{>}}C=O(g) + H_2(g)$

$\Delta H^{\ominus} = +84 \text{ kJ mol}^{-1}$

B $CH_3OH(g) + \frac{1}{2}O_2(g) \longrightarrow \overset{H}{\underset{H}{>}}C=O(g) + H_2O(g)$

$\Delta H^{\ominus} = -159 \text{ kJ mol}^{-1}$

(a)   Describe what is meant by *atom economy*.   [1]

(b)   Calculate the percentage atom economy for both reactions.   [2]

(c)   State which of reactions A and B is preferred, giving your reason.   [1]

(d)   A temperature of about 1000 K is used. What is the main operating concern?   [1]

(e)   High pressures are also a common feature of such reactions. Give one advantage and one disadvantage of using high pressures.   [2]

Interpretation

Questions on this part of the specification tend to be scattered throughout exam questions rather than asked in just one place. Sometimes the answers aren't always obvious and so familiarising yourself with lots of questions is useful. The answers do tend to follow certain themes – in this case the issues are linked to the efficient production of chemicals in an industrial process. Being able to complete yield and atom economy calculations is just one aspect. Being able to consider the operating conditions (temperature, pressure, catalyst, etc.) and safety is also important.

Ruairi's answer

(a) This is the amount of product made compared to the theoretical amount of product, expressed as a percentage. ✗

(b) Process A: AE = (30/32) × 100 = 94% ✓

Process B: AE = (30/48) × 100 = 63% ✓

(c) A is higher and so is more efficient. ✗

(d) High temperatures require a lot of energy and this can be costly. ✗

(e) Advantage: The reaction will be faster. ✓

Disadvantage: It will be very dangerous. ✗

MARKER COMMENTARY

(a) Ruairi has confused the definition of atom economy with yield. 0 marks.

(b) Fortunately he calculates it correctly and gains 2 marks. However, he has not used the A_r for hydrogen as given on the WJEC periodic table. It does not affect his calculation here but he should be careful to use the true value just in case.

(c) Ruairi's answer is not specific enough to gain the mark.

(d) The question refers to an operating concern and energy costs are not the most obvious answer whereas safety is. 0 marks.

(e) Ruairi gets 1 mark for his advantage but again his disadvantage answer is not worded specifically enough – he needs to add in what the danger is, e.g. of explosions. 1 mark.

Overall Ruairi gains 3 out of 7. This is quite a simple question as long as you have seen similar ones (and their answers before). Practising questions like this helps you to 'know what the questions are looking for'. Learning some specific answer types will be very useful.

Jessica's answer

(a) % atom economy = (mass of wanted products/total mass of products) × 100 ✓

(b) Process A: AE = (30.02/32.04) × 100 = 93.7% ✓

Process B: AE = (30.02/48.04) × 100 = 62.5% ✓

(c) A is preferred because there is less waste of materials. ✓

(d) Both reactants and products are highly flammable and so at this high temperature fire hazards will be a big concern. ✓

(e) Advantage: Increased pressure will increase the rate of reaction. ✓

Disadvantage: High pressures require expensive equipment such as high-pressure valves. ✓

MARKER COMMENTARY

(a) Jessica gives the correct meaning. 1 mark.

(b) Jessica calculates the answers correctly. 2 marks.

(c) She has learned what the importance of atom economy means to industrial processes and so gains the mark here.

(d) Jessica is aware that safety is a key operating factor and spots the issue with these flammable chemicals. 1 mark.

(e) Jessica gains 2 marks by giving a correct advantage and by making sure the wording of her disadvantage is specific.

Overall Jessica gains 7 out of 7 marks. Jessica knows her definitions and is also familiar with these types of questions so she knows what the answers are 'looking for'.

Q&A 2 There are many issues surrounding the provision of energy for running industrial processes. Traditionally, fossil fuels have been the main source of energy. The number of coal-fired power stations used to generate electricity has greatly reduced over the last century.

(a) Write an equation for the complete combustion of coal, including state symbols. [1]

(b) Give the meaning of the term *carbon neutral*. [1]

(c) Explain why using coal-fired power stations cannot be considered as carbon neutral. [2]

(d) Name two environmental issues associated with the operation of these power stations and give an example of why these issues concern us in each case. [4]

(e) Give another reason why coal-fired power stations have been phased out. [1]

Interpretation

The theme of this question is all about issues linked to 'the energy problem' such as availability of raw materials, pollution and carbon neutrality. These are common concepts and so, again, it is worth flicking through past papers to find out what similar questions they have asked and looking at how the answers are worded.

Ruairi's answer

(a) $C(s) + O_2(g) \longrightarrow CO_2(g)$ ✓

(b) Chemical processes that don't lead to an increase in carbon dioxide levels are carbon neutral. ✓

(c) Burning coal just releases carbon dioxide and it doesn't take any carbon dioxide in. ✓ ✗

(d) Burning coal can release sulfur dioxide which can cause acid rain. ✓ ✗

Burning coal releases lots of carbon dioxide which is causing global warming. ✓ ✗

(e) Coal is non-renewable. ✓

MARKER COMMENTARY

(a) Ruairi gives the correct equation. 1 mark.

(b) Ruairi has been given BOD (benefit of the doubt) – his answer is close and would have been perfect if he had referred to the atmosphere. 1 mark.

(c) Ruairi only gives half of the answer. He refers to the release of carbon dioxide but not to any processes where carbon dioxide is taken in – something that should always be considered in this type of question. 1 mark.

(d) Ruairi gives the names of both environmental issues but does not give examples of consequences. Poor answering technique so only 2 marks.

(e) Ruairi correctly identifies non-renewable as being the issue under question. 1 mark.

Overall Ruairi gains 6 out of 9 marks. More careful reading of questions and making sure answers are worded more closely to cover what the questions ask, will help Ruairi gain more marks.

Jessica's answer

(a) $C(s) + O_2(g) \longrightarrow CO_2(g)$ ✓

(b) A process is carbon neutral if there is no net increase in carbon dioxide levels in the atmosphere. ✓

(c) Burning coal releases carbon dioxide to the atmosphere. Although this carbon dioxide was originally removed from the atmosphere to make the trees to make the coal, it was over 500 million years ago. Using these power stations is a much shorter timescale process and so cannot be considered to be carbon neutral. ✓ ✓

(d) Acid rain which causes damage to trees and buildings made of rocks like limestone. ✓ ✓

Global warming which is causing climate change and things like rising sea levels and ice caps melting. ✓ ✓

(e) Coal is non-renewable. ✓

MARKER COMMENTARY

(a) Jessica gives the correct equation. 1 mark.

(b) Jessica gives an acceptable meaning. 1 mark.

(c) She answers this question well because she refers to the taking in and the release of carbon dioxide. She is told the process is not carbon neutral and so she deduces that timescale involved is important as well. 2 marks.

(d) Jessica correctly supplies the issues and examples of consequences. 4 marks.

(e) The issue is about coal being non-renewable and so Jessica gains 1 mark.

Overall Jessica gains 9 out of 9 marks. Through careful reading and precise answering, Jessica scores highly on this question.

2.4: Organic compounds

Topic summary

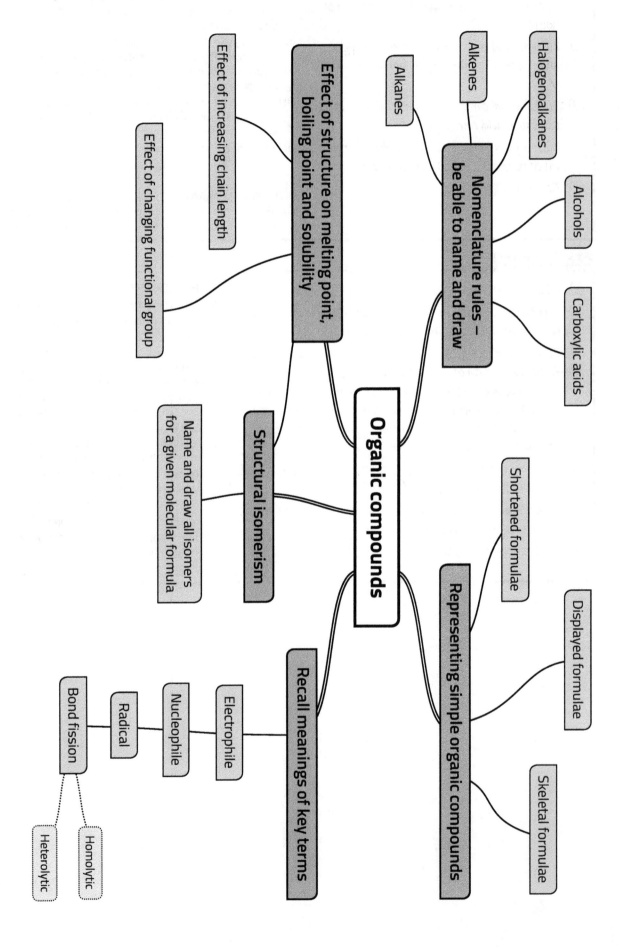

Q1 (a) What is meant by the term *structural isomerism*? [2]

....

....

(b) Illustrate your answer by drawing all the **structural** isomers of C_3H_5Cl. [4]

Q2 (a) Draw the displayed formula of *E*-but-2-ene. [1]

(b) Draw the displayed formula of an isomer of *E*-but-2-ene. [1]

(c) What is the empirical formula of *E*-but-2-ene? [1]

....

Q3 An unknown organic compound has the empirical formula CH_3O and an M_r of 62.

(a) Work out the molecular formula of this compound. [2]

....

....

(b) Draw the displayed formula of this compound given that it is an alcohol. [2]

(c) Name this compound. [1]

....

Q4 The table shows compounds which are all isomers with the molecular formula C_5H_{12}.

Isomer	Boiling temperature / °C
$(CH_3)_2CHCH_2CH_3$	27.8
$CH_3C(CH_3)_2CH_3$	9.5
$CH_3CH_2CH_2CH_2CH_3$	36.1

(a) Name the three compounds. [3]

....

....

....

(b) State and explain the trend in boiling temperature. [3]

....

....

....

....

Q5 Draw the skeletal formula of the compound 2-chloro-3-hydroxy-4-methylpentanoic acid. [1]

Q6 The diagram shows the structure of an unusual compound called oxacyclobutane.

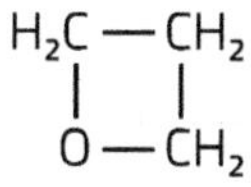

$$H_2C - CH_2$$
$$| \quad |$$
$$O - CH_2$$

(a) State the empirical formula of this compound. [1]

....

(b) Draw the structure of an isomer that does not contain a double bond. [1]

Q7 Analysis of an unknown organic compound X showed that it contained 18.60% carbon, 1.57% hydrogen, 55.03% chlorine by mass as well as oxygen. Addition of a solution of sodium hydrogencarbonate to X resulted in fizzing and X was neutralised by sodium hydroxide in a 1:1 ratio.

(a) Determine the empirical formula of compound X. [3]

(b) Explain what can be deduced from the observation with sodium hydrogencarbonate. [1]

....

....

(c) Explain what can be deduced from the observation with sodium hydroxide. [1]

....

....

(d) Use this information to identify the simplest compound that X could be. Explain your reasoning thoroughly. [3]

....

....

....

....

....

....

Question and mock answer analysis

Q&A 1   An unknown hydrocarbon has the molecular formula C_4H_8.

(a)   There are many isomers with this molecular formula.

   (i)   Draw and name the two straight chain isomers.   [4]

   (ii)   Draw and name a branched isomer.   [2]

(b)   The boiling point of such hydrocarbons is affected by their structure.

   (i)   Identify the hydrocarbon from part (a) with the lowest boiling temperature.   [1]

   (ii)   Explain why the boiling temperature is lower.   [2]

Interpretation

Structures and names of organic compounds were introduced at GCSE and you will have met hydrocarbons with structures up to four or five carbons in length. You will also have come across some simple names. At A-level you expand this knowledge significantly and you need to learn the stems, prefixes and suffixes for all the homologous series mentioned in the content map.

Part (b) requires recall from Unit 1 Section 1.4. This is a common question format and so you will need knowledge of the different types of intermolecular force and how and why they affect boiling and melting temperatures of organic compounds.

Ruairi's answer

(a) (i) but–1–ene ✓

but–2–ene ✗

(ii) 2–methylpropene ✓

(b) (i) 2–methylpropene ✓

   (ii) Methylpropene has weaker intermolecular forces, and so less energy is needed to boil methylpropene. ✗ ✗

MARKER COMMENTARY

(a)(i) The correct formulae and the first name each gain the marks. 3 marks.

   Ruairi has forgotten that some alkenes exist as *E-Z* isomers and so the correct letter must be included in the name – in this case *Z*-but-2-ene (either isomer is correct).

(ii) The correct formula and name each gain the mark. 2 marks.

In this case Ruairi is not penalised for including the 2 at the start of the name but it is not needed and so is best left out.

(b) (i) The correct compound is identified (name or structure will be accepted). 1 mark.

(ii) Ruairi does not state the name of the relevant intermolecular force and does not explain why it is weaker in the branched compound. He does, however, finish his answer with a reference to the amount of energy needed – this is sometimes awarded a mark. Always include this final phrase when the question refers to temperature (melting or boiling). 0 marks.

Overall Ruairi gains 6 out of 9 marks as he can write correct formulae and can name simple compounds. He has learned that structure affects boiling temperature but he has not focused on the precise wording required to gain marks when answering this type of question. His answers at first glance are promising and it is just that extra attention to detail that is missing. Practising questions like this will help him improve.

Jessica's answer

(a) (i)

Z-but-2-ene ✓   but-1-ene ✓

(ii)

2-methylpropene ✓

(b) (i) Methylpropene ✓

(ii) All the isomers have Van der Waals forces between their molecules but they are weakest in methylpropene because there is least surface contact between the molecules, and so less energy is needed to separate the methylpropene molecules. ✓✓

MARKER COMMENTARY

(a) (i) The correct formula and name each gain the mark. 4 marks.

(ii) The correct formula and name each gain the mark. 2 marks.

The 2 at the start of the name is not required as there is nowhere else for the methyl group to go.

(b) (i) The correct compound is identified (name or structure will be accepted). 1 mark.

(ii) Jessica correctly names the type of intermolecular force in each compound and then explains why it is weaker in the branched compound. 2 marks

Overall Jessica gains 9 out of 9 marks as all the content is correct. Jessica has worded her answer to (b)(ii) very well. Many students struggle to write correctly worded explanations to structure and bonding questions and so it is worth learning good answers like this 'off by heart'.

Q&A 2   An unknown organic compound is found to contain 52.1% carbon, 34.7% oxygen and hydrogen.

(a) Given that it has an M_r of 46.06, use this data to prove that the molecular formula is C_2H_6O.   [3]

(b) Draw the displayed formulae of the two isomers of this molecular formulae.   [2]

(c) Identify which isomer is more soluble in water. Explain your answer using a labelled diagram to help.   [4]

Interpretation

Empirical formula calculations are taught in Section 1.3 but can be assessed in any part of the A level course. They are common in Unit 2 organic chemistry questions.

It is common to be asked to draw isomers – make sure you check whether it is a structural, displayed or skeletal formula that is required.

Part (c) requires recall from Unit 1 Section 1.4. This is a common question format and so you will need knowledge of the different types of intermolecular force and how and why they affect boiling, melting temperatures and (in this case) solubility of organic compounds.

Ruairi's answer

(a) Percentage of hydrogen = 13.2% ✓

$C = 52.1 / 12 = 4.342$

$H = 13.2 / 1.01 = 13.06$ ✓

$O = 34.7 / 16 = 2.169$

Simplest ratio = 2:6:1 i.e. C_2H_6O

Therefore the compound has the formula C_2H_6O ✗

(b)

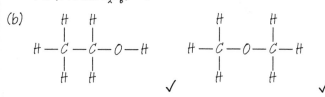

H — C — C — OH   ✗

H — C — O — C — H   ✓

(c) Ethanol ✓

There is hydrogen bonding in ethanol and this makes it water soluble.

hydrogen bond

✓ ✗ ✗ (It is possible that this explanation would not even score 1 mark on some mark schemes.)

MARKER COMMENTARY

(a) Ruairi has shown he can correctly calculate the empirical formula but has failed to answer the last part of the question where he needed to show that the empirical formula was also the molecular formula for this example. Following his calculation route, he needed to finish his answer by calculating the mass of the empirical formula unit and compare it to the M_r given in the question. 2 marks.

(b) Ruairi has forgotten to show the OH bond. 1 mark.

(c) This answer reflects a lack of technique and lack of knowledge. Ruairi needs to recall the definition of hydrogen bonding (from Unit 1 Section 1.4) more accurately and make sure he always shows the polarisation of the O–H bond and the lone pair that forms the hydrogen bond with a δ^+ hydrogen atom on an adjacent molecule. When answering questions about solubility in water the hydrogen bonds must always be drawn between the organic molecule and a water molecule. Finally, when the question is about a comparison between two substances, the answer should refer to both substances and not just one of them. Ruairi does not mention the other isomer at all. 2 marks.

Overall Ruairi gains 5 out of 9 marks as careless errors and lack of recall have lost him marks. Plus, he has not always answered all parts of the question – he needs to read questions more carefully.

Jessica's answer

(a) Percentage of hydrogen = 13.2% ✓

$C = (52.1 / 12) \times 46.06 = 200$

$H = (13.2 / 1.01) \times 46.06 = 602$ ✓

$O = (34.7 / 16) \times 46.06 = 99.9$

Integer ratio = 2:6:1 therefore the compound has the formula C_2H_6O ✓

(b)

H — C — C — O — H   ✓

H — C — O — C — H   ✓

(c) Ethanol ✓

Ethanol can form hydrogen bonds with water.

hydrogen bond

The other isomer cannot form hydrogen bonds with water.

✓ ✓ ✓

MARKER COMMENTARY

(a) Jessica has answered this question well because she has shown how to use the data to determine the molecular formula not just the empirical formula. Then she has linked her answer back to the question being asked 'use this data to prove' – this is good technique. 3 marks.

(b) Jessica has correctly drawn both isomers showing all the bonds. 2 marks.

(c) Jessica has correctly identified that ethanol is more soluble in water. She has good recall of the definition of hydrogen bonding and so includes all the required detail in her diagram:
- The O–H bonds in water and ethanol are very polarised.
- Hydrogen bonds form between the lone pairs on the oxygen atoms of ethanol and the δ^+ hydrogen atoms on adjacent water molecules (or vice versa).

She finishes her answer by including why the other isomer is less soluble. 4 marks.

Overall Jessica gains 9 out of 9 marks as she recalls all the factual content correctly and is able to apply these ideas.

2.5: Hydrocarbons

Topic summary

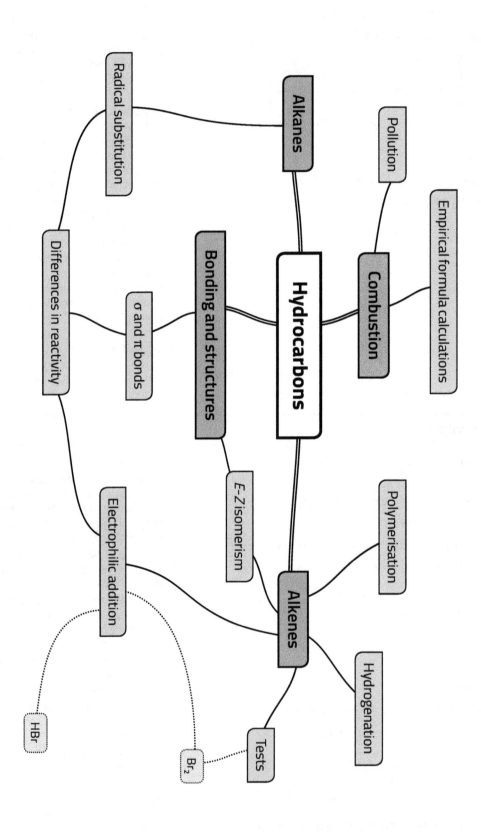

Q1   The combustion of fuels obtained from crude oil is a topic causing much concern at current times.

(a)   Name the process by which petrol is obtained from crude oil.   [1]

....

(b)   The composition of petrol varies greatly but one of its commonest components is octane. Draw the displayed formula of octane.   [1]

(c)   Another component is commonly called isooctane and it has the structure shown below. Give the IUPAC name for isooctane.   [1]

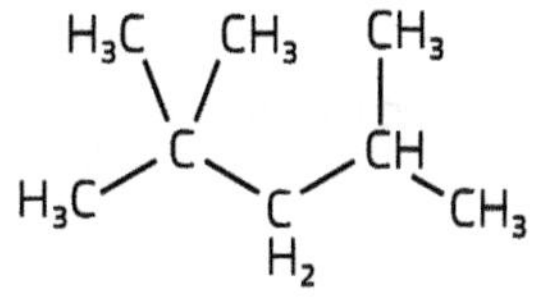

....

(d)   Explain why isooctane has a lower boiling temperature than octane.   [2]

....

....

....

(e)   Write an equation for the incomplete combustion of octane to form carbon monoxide.   [2]

....

(f)   Why is there concern over this emission from cars?   [1]

....

Q2 Compound A is made by the reaction of an alkane with chlorine. Analysis of compound A showed that it was made up of 37.8% carbon, 6.3% hydrogen and 55.9% chlorine.

(a) Work out the empirical formula of A. [2]

(b) Given that A has a M_r of 127, work out the molecular formula of A. [1]

....

(c) Name the alkane from which A was made. [1]

....

(d) Name the reaction by which A was formed. [1]

....

(e) 1-chloropropane is formed by the same type of reaction as A. Outline the mechanism by which 1-chloropropane is formed. [4]

(f) State and explain two reasons why other products besides 1-chloropropane will form. [4]

Q3 (a) Describe the nature of bonding in alkenes and explain how this can lead to the existence of E-Z isomers for some alkenes. [5]

(b) Draw and label the E-Z isomers of 2-chloro-3-methylpent-2-ene. [2]

Q4 Pent-1-ene reacts with hydrogen bromide to give 2-bromopentane as the major product.

(a) Name the mechanism for this reaction. [1]

(b) Draw the mechanism for this reaction. [3]

(c) Explain why 2-bromopentane is the major product of this reaction. [2]

....

....

....

(d) But-1-ene can be polymerised.

(i) Name the type of polymerisation. [1]

....

(ii) Draw the section of polymer made from 3 monomer units, when but-1-ene is polymerised. [2]

Q5 An unknown hydrocarbon B was reacted with oxygen at 1 atm and 373 K. Complete combustion of 5.00×10^{-2} g of this hydrocarbon produced 36.5 cm^3 of carbon dioxide.

(a) Calculate the M_r of the unknown hydrocarbon. [3]

(b) Deduce the molecular formula of B – show your working out. [2]

(c) Given that B does not decolourise bromine water, suggest a structure for B. [1]

Q6 (a) An alkene reacts with hydrogen in a 1:3 ratio. What does this tell you about the alkene? [1]

....

(b) (i) Write an equation for the reaction between but-2-ene and hydrogen. [1]

....

(ii) State and explain, out of the product and but-2-ene, which, if any, will have the higher boiling temperature. [3]

....

....

....

(c) But-2-ene also reacts with hydrogen bromide.

(i) Write an equation for this reaction. [1]

....

(ii) Many reactions of alkenes with hydrogen bromide produce two products with one product being produced in far greater quantities than the other. Explain why the reaction with but-2-ene only produces one product. [2]

....

....

....

Question and mock answer analysis

Q&A 1 (a) Describe in detail the similarities and differences between the nature of bonding in simple alkanes and alkenes. [4]

(b) Explain why some alkenes display *E-Z* isomerism. [2]

(c) A student wants to find out how many double bonds there are in 0.2 mol of an unknown alkene. They decide to carry out a titration using bromine in a burette. After carefully adding bromine to their sample, they determined that 30.75 cm^3 of bromine reacted with the alkene. Use the data to show that there are three double bonds in the alkene. Explain your reasoning fully, including an explanation as to how the endpoint was obtained. The density of bromine is 3.119 g/cm^3. [4]

Interpretation

Alkanes and alkenes have different reactivities linked to the bond types they contain. It is important that you know the definitions of each bond type and how they affect reactivity for all functional groups at AS level.

The calculation is a familiar calculation in an unfamiliar context. Practising questions like this, especially ones involving rearranging the equation for density, is useful because you will become more adept at applying knowledge to unfamiliar situations the more practice you do.

Ruairi's answer

(a) Alkanes are made up of only single bonds between all the atoms. Alkenes can have double bonds as well between two carbon atoms, and this makes alkenes more reactive. ✓ X X X

(b) Alkenes cannot rotate about the double bond and so this means two structures can form which are isomers. ✓ X

(c) Mass of bromine = density × volume = 95.91g ✓

Moles of bromine = 95.91/79.9 = 1.2 X

Ratio 0.2 to 1.2 = 1 to 6

Because 2 moles of bromine react for every double bond the alkene has 3 double bonds. X

MARKER COMMENTARY

(a) Ruairi has not recognised the detail required by the question and he does not have full recall of the different types of bonding in hydrocarbons. 1 mark is awarded for recognising that alkanes do not have double bonds.

(b) Ruairi is aware that double bonds affect the structure of alkenes, but he does not know the full definition for why alkenes can have *E-Z* isomers. Only 1 mark.

(c) Ruairi lost 3 marks because he initially made a mistake with the M_r of bromine, and then he made up an incorrect ratio in order to fit his answer to the one supplied.

Overall Ruairi gains 3 out of 10 marks. He has struggled with this question. Lack of thorough recall affected his first answers. He needs to spend more time learning facts about bonding and isomerism.

In part (c) he was not really sure what was going on. If he had written an equation for an alkene reacting with bromine, he might have recognised his mistake and fixed his answer. In unfamiliar situations start by linking it to facts you know to help you visualise what is going on.

Jessica's answer

(a) Alkanes and alkenes both contain σ-bonds. These form between C and C atoms and between C and H atoms by the 'head on' overlap of s-orbitals with other s or p-orbitals. Only alkenes contain π-bonds, which occur between C atoms. π-bonds are formed by the sideways overlap of p-orbitals. ✓✓✓✓

(b) Alkenes exhibit E-Z isomerism because there is restricted rotation about the double bond. This means that if a molecule has two different groups attached to each end of the double bond, they can be in two different orientations with respect to each other and so two isomers are possible. ✓✓

(c) The test for a double bond is to add bromine or bromine water which is decolourised. This colour change can be used as the endpoint for the titration. ✓

Mass Br_2 = density × volume = 95.91g ✓

Moles Br_2 = 95.91/159.8 = 0.600

1mole of Br_2 reacts with one double bond ✓

Ratio of Br_2 to alkene is 1 to 3

So there must be 3 double bonds in the alkene – as said in the question. ✓

MARKER COMMENTARY

(a) Jessica has revised thoroughly and knows what σ- and π-bonds are. She also makes sure she talks about both alkanes and alkenes in her answer to ensure she gives all the detail required.

(b) Jessica has seen this question before, and her answer is precise. She uses the phrase 'restricted rotation' rather than 'no rotation' plus she recalls the exact condition for some alkenes to exist as isomers – she has learned from reading mark schemes.

(c) Jessica realises that a colour change is needed for the endpoint of a titration and she links this to her knowledge of the test for C=C bonds in alkenes. She knows that bromine and C=C react in a 1 to 1 ratio. She then completes the calculation correctly and is happy that her answer agrees with the one given.

Overall Jessica gains 10 out of 10 marks. Thorough recall of definitions helps her to answer the first two parts correctly. Good knowledge of the reactions of alkenes helps her to understand the principles behind the question in part (c).

Q&A 2

Alkanes can be converted into halogenoalkanes. The reaction is called radical substitution.

(a) What is meant by the terms 'radical' and 'substitution' in this context? [2]

(b) What condition is essential for this reaction to occur? [1]

(c) Write an equation for the formation of dibromomethane from methane. [1]

(d) Dibromomethane is a colourless liquid and can be used as an organic solvent. What mass of bromine would be required to produce 2.0 kg of dibromomethane? Give your answer to the appropriate number of significant figures. [3]

(e) The reaction occurs in steps, just like photochlorination of methane. Write a propagation equation that forms dibromomethane. [2]

Interpretation

This question is about the reaction of alkanes with halogens. It is important that all the key terms are memorised as they often give useful points in questions, and they also help you understand – if you don't know what the words mean you won't understand the question.

Photochlorination is studied during AS but the reaction can be applied directly to other halogens. The same issues also apply. All you have to do is remember the reaction sequence with chlorine but swap bromine in instead.

Ruairi's answer

(a) Radical – something with one electron.
Substitution – where something is swapped for something else. ✗ ✓

(b) UV light ✓

(c) $CH_4 + Br_2 \longrightarrow CH_2Br_2 + H_2$ ✗

(d) M_r of $CH_2Br_2 = 173.82$
Moles of dibromomethane = 2000/173.82
= 11.5 ✓
Mass of bromine = 11.5 × 159.8 = 1837g ✓ ✗

(e) $CH_3Br + Br^• \longrightarrow {}^•CH_2Br + HBr$ ✓ ✗

MARKER COMMENTARY

(a) Ruairi has some recall of key terms but his wording for the meaning of 'radical' is too ambiguous – most radicals will have more than one electron and the important aspect is that it has an **unpaired** electron. So only 1 mark.

(b) Ruairi has good recall of the conditions for this reaction. 1 mark.

(c) Ruairi has forgotten that the other product must be a hydrogen halide and so he creates an incorrect equation. 0 marks.

(d) Ruairi is able to complete mass calculations but has failed to report his answer to 2 sf and so loses a mark. 2 marks.

(e) Ruairi has not given an equation that produces dibromomethane but because he has given a correct propagation equation he does actually get awarded 1 of the marks.

Overall Ruairi only gains 5 out of 9 marks. He could easily improve this score by spending time learning key terms more carefully so that he doesn't throw away marks for poor wording. He also needs to ensure he reads the question more carefully, so he spots a comment like 'the appropriate number of significant figures' and doesn't lose marks for avoidable errors.

Jessica's answer

(a) Radical – a species with an unpaired electron.
Substitution – where one atom or group replaces another atom or group. ✓ ✓

(b) UV light ✓

(c) $CH_4 + 2Br_2 \longrightarrow CH_2Br_2 + 2HBr$ ✓

(d) No. of moles of CH_2Br_2 = 2000/173.82 = 11.5 ✓
Mass of Br_2 = 11.5 × 159.8 = 1837g ✓
= 1.8kg (2 sf) ✓

(e) ${}^•CH_2Br + Br_2 \longrightarrow CH_2Br_2 + Br^•$ ✓ ✓

MARKER COMMENTARY

(a) Jessica has learned her key terms and gets 2 marks.

(b) Jessica has good recall of the conditions for this reaction. 1 mark.

(c) Jessica has remembered that hydrogen halides are always produced and so she knows that she needs 2 moles of bromine to balance this equation. 1 mark.

(d) Jessica can complete mass calculations correctly and she has noted that the answer should be reported to 2 sf (in keeping with 2.0 kg). 3 marks.

(e) Jessica has recognised that she needs to have an equation where a radical reacts with a molecule to create a new radical and a new molecule (namely dibromomethane) and she has worked out the correct equation to use. 2 marks.

Overall Jessica gains 9 out of 9 marks. Thorough recall of definitions helps her to answer the first three parts correctly. Good maths skills and careful reading of the question (so she noticed the comment about significant figures) ensures she gains full marks on the calculation. The last question is tricky but because she understands what a propagation reaction is, and she knows what product to make, she is able to construct the correct equation. The more you practise questions like this, the easier they become.

2.6: Halogenoalkanes

Topic summary

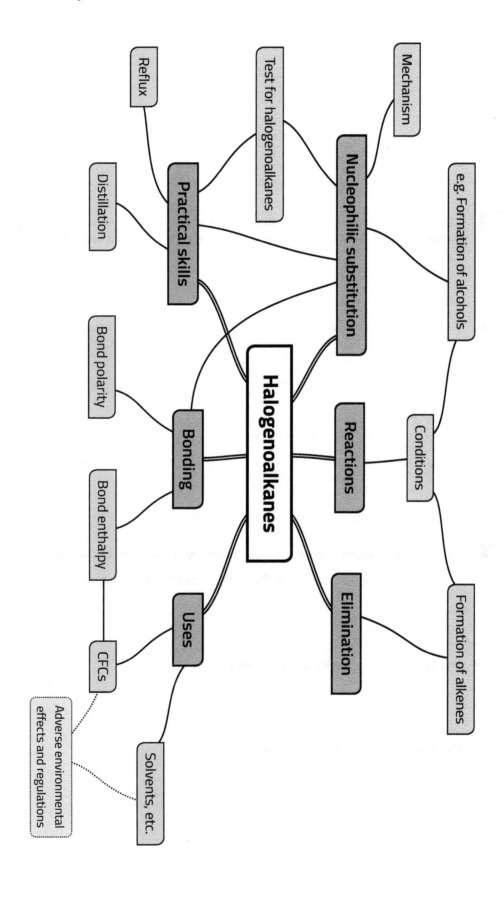

Q1 Explain what is meant by nucleophilic substitution in halogenoalkanes and explain why they undergo this reaction. [4]

....

....

....

....

....

....

....

Q2 Explain in detail (including equations) why the use of CFCs was restricted following the Montreal protocol in 1987. What other type of chemicals did scientists then recommend and why? [6]

....

....

....

....

....

....

....

....

....

....

Q3 Halogenoalkanes can be converted into alcohols by nucleophilic substitution reactions.

(a) Give the name of a halogenoalkane that could be used to make butan-2-ol. [1]

....

(b) State the reagent and solvent required for this reaction. [2]

....

(c) Give the chemical equation for this reaction. [1]

....

(d) The initial reaction requires refluxing of the chemicals but then the product is isolated by distillation. Draw the apparatus for distillation. [3]

(e) One student did not think that the product could be separated from the reactant by distillation. Explain why they thought this and suggest a reason why the separation would actually work. [4]

Q4 A student had three bottles each containing a sample of a halogenobutane derivative – one with chlorine, one with bromine and one with iodine as the halogen substituent.

(a) Outline the test(s) that could be carried out to unambiguously identify which bottle contained which sample – giving appropriate reagents, observations and ionic equations for the precipitation reactions. [6]

(b) If the only difference between the halogenoalkanes mentioned in part (a) was the halogen, what would be the order of hydrolysis? Explain your answer. [3]

(c) Suggest the order of hydrolysis if the samples had different structures – one was a tertiary halogenoalkane, one was a secondary and one was a primary. Give a reason for your thinking. [2]

Q5 Halogenoalkanes can also be converted into alkenes.

(a) Give the name for this type of reaction. [1]

(b) Give the name of a halogenoalkane that could be used to make hex-3-ene. [1]

(c) State the reagent and solvent required for this reaction. [2]

(d) Give the chemical equation for this reaction. [1]

(e) Multiple products are possible with such a reaction.

(i) State a product that is a structural isomer of hex-3-ene and explain how it could be produced. [2]

....

....

....

(ii) Draw the two *E–Z* isomers of hex-3-ene and label them. [2]

(iii) State a product that has a different functional group to hex-3-ene and explain how it could be produced. [2]

....

....

....

Q6 There are very many halogenoalkanes with very many uses.

(a) Draw the displayed formula of 2,2,3-trifluoro-1-iodopentane. [1]

(b) Draw the skeletal structure of the compound made when 2-chloro-4-methyloctane is hydrolysed. [1]

(c) Freon-12 was once used as a refrigerant. Its structure is shown below:

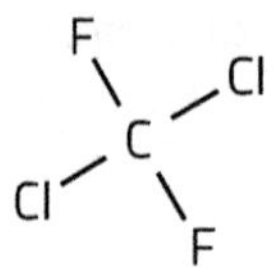

$$
\begin{array}{c}
F \\
\diagdown \\
Cl \diagup C \diagup Cl \\
\diagup \diagdown \\
 F
\end{array}
$$

(i) Give the IUPAC name of Freon-12. [1]

....

(ii) State a property that made it useful as a refrigerant. [1]

....

(iii) Freon-12 was initially replaced by $HCClF_2$, which had a lower ozone depletion effect. Draw the structure of a similar molecule that should have no ozone depletion affect. [1]

(d) State another use of halogenoalkanes. [1]

....

Q7 Smaller halogenoalkanes are volatile substances and also virtually insoluble in water.

(a) Explain why 2-chloropropane has a lower boiling temperature than 1-chloropropane. [3]

....

....

....

....

(b) Explain why 1-chloropropane is insoluble in water. [1]

....

....

Question and mock answer analysis

Q&A 1 Chloroethane has been widely used commercially. Nowadays its use as a refrigerant is restricted.

(a) Draw the skeletal structure of chloroethane. [1]

(b) Chloroethane can be converted into ethanol.

  (i) Write an equation for this reaction. [1]

  (ii) A student stated that distillation could be used to separate the product. Explain, in detail, why the student was incorrect. [2]

(c) A refrigerator contains 1.00 kg of chloroethane.

  (i) Calculate the number of molecules of chloroethane in the refrigerator. Give your answer to the appropriate number of significant figures. [3]

  (ii) Calculate the volume of gas released if all this refrigerant escaped to the environment. Assume that the release took place at room temperature under standard conditions. [2]

  (iii) Why would the release of this gas be a cause for concern? [1]

Unit 2 Practice questions

Interpretation

Representing organic compounds in all the forms covered in Section 2.4 of the specification is a skill that is applicable to all areas of organic chemistry. It is important that you know how to do each type of structure representation.

Other transferable skills, such as the calculations met in Section 1.3 are also required for completing organic papers as there is a quota for 'maths marks' in all unit papers. Again, make sure you practise calculations such as the ones shown here.

Ruairi's answer

(a) $H_3C \frown Cl$ ✗

(b) (i) $C_2H_5Cl + NaOH \longrightarrow C_2H_5OH + NaCl$ ✓

  (ii) Chloroethane is a gas and so would distil over first. ✓✗

(c) (i) Number of moles = 1000/64.55 = 15.5

  Number of molecules = 15.5 × 6.02 × 10^{23} = 9.33 × 10^{24} ✓✓✓

  (ii) PV = nRT

  V = nRT/P = (15.5 × 8.31 × 298)/1

  = 38,383m³ ✓✗

  (iii) CFCs damage the ozone layer. ✗

MARKER COMMENTARY

(a) Ruairi has made a common mistake of showing a methyl group at the end of a skeletal structure. 0 marks.

(b) (i) Ruairi gains this mark and is not penalised for abbreviating the structure of the alkyl chain because the structure is unambiguous in this case. 1 mark.

(ii) Ruairi gains 1 mark for giving some correct information relating to the different boiling temperatures and separation of the chemicals involved. He does not gain two marks, however, because he does not go into sufficient detail to explain why the student was incorrect.

(c) (i) Ruairi also completes the calculation correctly and gives his answer to 3 significant figures. He rounds his answer as he goes along. Fortunately, this does not affect this question but it's not a good idea as it can lead to an answer outside the acceptable range. 3 marks.

(ii) Ruairi uses the ideal gas equation, which would have worked if he had converted his pressure to pascals. He gets one mark for correct use of this equation and he gives the volume in the units correct for this calculation. 1 mark.

(iii) Ruairi has learned about the risks of halogenoalkanes and is correct in his statement about CFCs, but chloroethane is not a CFC and this means he gains no marks.

Overall Ruairi gains 6 out of 10 marks. He has made a few minor mistakes – with an incorrect skeletal structure and using the wrong value for pressure. These can be corrected with practice. He needs to work on the wording of his answers – taking more care to make the meanings chemically correct and clear.

Jessica's answer

(a) [skeletal structure with Cl] ✓

(b) (i) $CH_3CH_2Cl + NaOH \rightarrow CH_3CH_2OH + NaCl$ ✓

(ii) Ethanol has a higher boiling point, than chloroethane, due to hydrogen bonding between the molecules. ✓

In distillation it is the chemical with the lower boiling point that distils over first and so ethanol would not be separated. ✓

(c) (i) Number of moles = 1000/64.55 = 15.49

Number of molecules = 15.49 × 6.02 × 10^{23}
= 9.33 × 10^{24} ✓✓✓

(ii) V = 24.5 × 15.49 = 380dm³ ✓✓

(iii) Chloroalkanes would produce chlorine radicals in the upper atmosphere and thus damage the ozone layer. ✓

MARKER COMMENTARY

(a) Jessica can draw skeletal structures correctly. 1 mark.

(b) (i) Jessica gives the correct equation and shows good technique by giving the structural formula of both organic chemicals. 1 mark.

(ii) Jessica gives a detailed answer which explains the differences in boiling temperatures and how this affects the separation of the product from the mixture. She clearly answers the question asked. 2 marks

(c) (i) Jessica completes the calculation correctly and gives her answer to 3 significant figures. She leaves rounding until the last step, which is always the best thing to do.

(ii) Jessica uses the gas molar volume for room temperature and 1 atm. She calculates the volume correctly and gives the correct units for this calculation. 2 marks.

(iii) Jessica has learned about the issues with halogenoalkanes creating chlorine radicals in the upper atmosphere and is aware that chloroethane is not a CFC. She words her answer appropriately. 2 marks.

Overall Jessica gains 10 out of 10. She is competent at drawing different structures and writing equations and she has good calculation skills. All these come with practice. She also reads the questions carefully and links the information to aspects of work she has learned in lessons correctly.

Q&A 2 Halogenoalkanes undergo both nucleophilic substitution reactions and elimination reactions with sodium hydroxide.

(a) What condition would favour elimination reactions? [1]

(b) Write an equation for the reaction of 2-bromo-2-methylbutane with sodium hydroxide under these conditions. [2]

(c) Draw the apparatus that could be used to separate the organic product from the reactants. [3]

(d) Give a reagent that could be used to test that the elimination reaction had worked. Explain your answer, including the result of the test. [2]

(e) Draw the structure of an isomer of 2-bromo-2-methylbutane that would not undergo an elimination reaction with sodium hydroxide. Explain your reasoning. [2]

Interpretation

To be competent at answering organic chemistry questions, a thorough recall of reactions, reagents and conditions across all the different functional groups is necessary. Make lists and learn them.

It is also important to understand why reactions happen – this could be due to polarisation from differences in electronegativity or, as in this case, due to how the atoms are arranged in the structure.

Ruairi's answer

(a) Heat ✗

(b) $CH_3C(CH_3)BrCH_2CH_3 + NaOH \longrightarrow CH_2C(CH_3)CH_2CH_3 + NaBr$ ✓✓

(c)

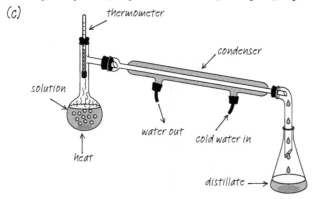

✓✓✗

(d) Add silver nitrate solution and a cream ppt would confirm the presence of bromide ions from the sodium bromide. ✗✗

(e)

There are no hydrogen atoms that can be removed from the carbon next to the bromine group and so HBr cannot be eliminated. ✗✓

MARKER COMMENTARY

(a) The key condition is the solvent and so Ruairi gains no marks for mentioning heat. 0 marks.

(b) Ruairi constructs a correct equation giving the other possible alkene product. Either is acceptable. 2 marks.

(c) Ruairi only gains two marks because he is not careful with his placement of the thermometer and it is too low. 2 marks.

(d) Whilst Ruairi is correct that silver nitrate would give a cream ppt because bromide ions are present – what he has overlooked is the fact that the halogenoalkane itself could hydrolyse to produce bromide ions. This test does not prove that an alkene is made. 0 marks.

(e) Ruairi understands the requirements for a halogenoalkane to undergo elimination. Unfortunately, he draws a 'correct' structure but includes an extra methyl group and so loses the isomer mark. Only 1 mark.

Overall Ruairi gains 5 out of 10 marks. Ruairi will easily gain more marks with a little more attention to detail. He should practise drawing diagrams. He needs to learn reagents and conditions for all the organic reactions. And he needs to read questions carefully and check that his answer complies with all parts.

Jessica's answer

(a) Pure ethanol. ✓

(b) $CH_3C(CH_3)BrCH_2CH_3 + NaOH \longrightarrow (CH_3)_2CCHCH_3 + NaBr$ ✓✓

(c)

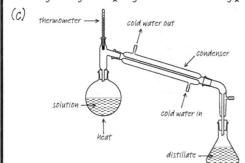

✓✓✓

(d) Shake the distillate with bromine water. It would decolourise because the bromine water reacts with the alkene produced. The reactant would not decolourise bromine water. ✓✓

(e)

$$H_3C - \underset{\underset{CH_3}{|}}{\overset{\overset{CH_3}{|}}{C}} - CH_2Br$$

There are no hydrogen atoms on the C atom adjacent to the C–Br bond and so elimination cannot occur. ✓✓

MARKER COMMENTARY

(a) Jessica knows the conditions needed to favour elimination over nucleophilic substitution – the absence of water. 1 mark.

(b) Jessica gives the correct structure for the halogenoalkane and a correct alkene that could be produced by elimination. And she completes the equation correctly. 2 marks.

(c) Jessica gains 3 marks for giving a diagram that 'works' – there is a heat source, flask and sideways condenser; the water goes in at the bottom and out at the top; the thermometer is opposite the exit to the condenser. 3 marks.

(d) Jessica correctly picks a test that proves that an alkene is made, and she even includes in her answer the fact that the reactant would not react to give a result with this test – good technique. 2 marks.

(e) Jessica understands the requirements for a halogenoalkane to undergo elimination and draws the correct isomer. 2 marks.

Overall Jessica gains 10 out of 10. She has good learning of reagents and reactions. She has learned the important aspects required when drawing organic apparatus set ups – reflux and distillation. She also knows how to word answers well (probably from reading mark schemes).

Unit 2 Practice questions

2.7: Alcohols and carboxylic acids

Topic summary

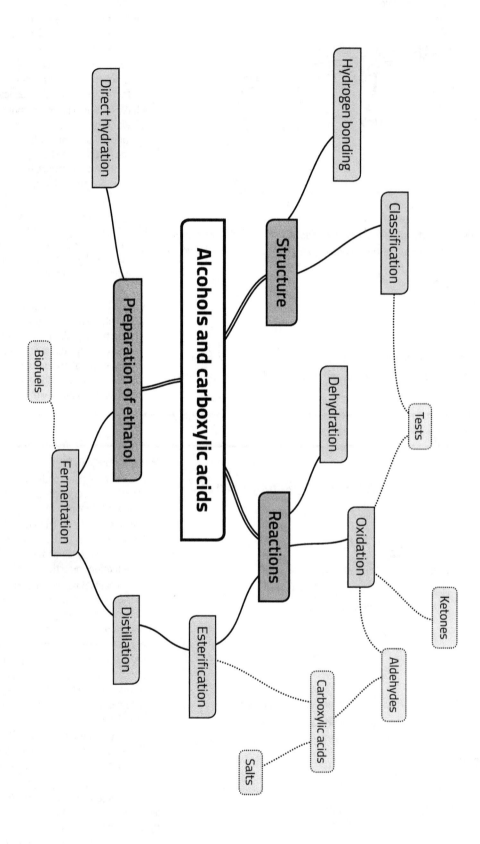

Q1 A student wanted to make a sample of propanal. Describe the procedure, including reagents, conditions and essential apparatus, to create and isolate this product. Include relevant equation(s) and diagrams in your answer. [6]

....

....

....

....

....

....

....

....

....

....

Q2 (a) Write an equation for the formation of butyl ethanoate, showing clearly the structure of the product. [2]

....

(b) Name the catalyst for this reaction. [1]

....

(c) When the reaction mixture is poured into a small beaker of water, the ester alone forms as a layer on the surface of the water. Explain why this occurs. [2]

(d) Name a use of butyl ethanoate. [1]

Q3 Ethanol is usually produced by fermentation of sugars, but it can be made from ethene.

(a) Name the type of reaction involved in the production of ethanol from ethene. [1]

(b) Write an equation for the formation of ethanol from ethene. [1]

(c) What conditions are required to make this reaction occur? [3]

(d) Why might this process be favoured over fermentation in certain countries? [1]

(e) What is meant by the term carbon neutral? [1]

(f) State two reasons why the formation of ethanol from ethene is not carbon neutral. [2]

Q4 Aldehydes and ketones both contain a C=O functional group.

(a) Name the reaction by which they are formed from alcohols. [1]

(b) What reagent(s) would be used for these conversions? [1]

(c) How would you know the reaction had occurred? [1]

....

(d) Write an equation for the formation of pentan-2-one. [1]

....

Q5 Explain why ethylamine ($CH_3CH_2NH_2$) is a gas at room temperature but ethanol (CH_3CH_2OH) is a liquid. [3]

....

....

....

....

....

....

Q6 (a) Write an equation for the formation of butanoic acid from the appropriate alcohol. [1]

....

(b) Draw the apparatus required to achieve this conversion and explain why it is used. [4]

....

....

....

....

(c) Butanoic acid reacts with alcohols to form esters. What alcohol would you need to make $CH_3CH_2CH_2COOCH(CH_3)_2$? [1]

....

(d) Butanoic acid also reacts with sodium carbonate. Write an equation for this reaction. [2]

(e) Outline a procedure for obtaining a pure, crystalline sample of the product. [3]

Q7 Alcohols are highly flammable chemicals.

(a) Write an equation for the complete combustion of propan-1-ol. [2]

(b) Complete combustion of 2.00 g of an unknown alcohol produces 4.75 g of carbon dioxide and 2.43 g of water. Use this information to calculate the molecular formula of the alcohol. [3]

(c) Ethanol, produced by fermentation of sugar cane, is considered to be a biofuel.

(i) What is meant by the term *biofuel*? [1]

(ii) Give one advantage and one disadvantage of using ethanol as a biofuel. [2]

Q8 The oxidation of alcohols is a very useful reaction.

(a) Name the alcohol needed to produce butanal. [1]

....

(b) Name the product formed by the complete oxidation of pentan-3-ol. [1]

....

(c) Draw the skeletal structure of the product formed by the complete oxidation of 3-chloro-2-methylpropan-1-ol. [2]

(d) Draw and name an isomer with the same molecular formula as 3-chloro-2-methylpropan-1-ol that cannot be oxidised. [2]

....

Q9 Fermentation of sugars is an important process for producing ethanol for use as a biofuel.

(a) Write the equation for this reaction. [1]

....

(b) Under certain conditions producing ethanol by fermentation can be considered to be carbon neutral. Referring to the key stages in this process, explain how this can be true. [4]

....

....

....

....

....

....

Question and mock answer analysis

Q&A 1 Ethanol can be prepared by the fermentation of glucose ($C_6H_{12}O_6$).

(a) (i) Write an equation for this reaction. [2]

(ii) State the essential conditions for this reaction to occur. [3]

(b) Name the alternative route to produce ethanol from ethene and state why this route is considered to be less 'green' than fermentation. [2]

(c) (i) Ethanol produced by fermentation is considered to be a biofuel. What is meant by the term *biofuel*? [1]

(ii) Give a disadvantage of biofuels. [1]

Interpretation

This is a straightforward question as most of it comes under the AO1 category in the mark scheme – the part that links to recall of knowledge. In exam papers about a third of the paper is pure recall. Many questions start with parts that require good recall and then they develop and expect you to apply your knowledge – so it is important to learn all the processes such as the two mentioned here in order that you can get the first 'easy' marks on exam questions.

Ruairi's answer

(a) (i) $C_6H_{12}O_6 \rightarrow 2C_2H_5OH + 2CO_2$ ✓✓

(ii) Yeast and at body temperature. ✓✓✗

(b) Hydrolysis. ✗

This process requires high temperatures which are created using fossil fuels. ✗

(c) (i) A fuel that is made from plant matter. ✓

(ii) Growing crops for biofuels needs large quantities of water. ✓

MARKER COMMENTARY

(a) (i)The correct products and balancing each gain the mark. 2 marks.

(ii) Ruairi fails to notice that the question is worth 3 marks and whilst his recall is good, he fails to gain the third mark. Only 2 marks awarded.

(b) Ruairi confuses the addition of water (hydration) with the splitting by water (hydrolysis) and so loses the first mark. He does recall a valid reason why hydration is less 'green' but poor technique means that he did not mention the other process in his answer. Comparison answers are common and require students to mention both processes in an answer. 0 marks.

(c) (i) Correct definition. 1 mark.

(ii) Correct disadvantage of biofuel. 1 mark.

Overall Ruairi gains 6 out of 9 marks. His recall is quite good, apart from muddling hydrolysis and hydration, but poor question technique led to loss of 2 marks. With a more careful approach to reading and answering he can learn to avoid these mistakes.

Jessica's answer

(a) (i) $C_6H_{12}O_6 \rightarrow 2C_2H_5OH + 2CO_2$ ✓✓

(ii) Yeast and a temperature of about 30°C, in the absence of air. ✓✓✓

(b) Hydration. ✓

Ethene is obtained from a non-renewable source whereas glucose is renewable. ✓

(c) (i) A biofuel is a fuel that has been produced using a biological source. ✓

(ii) Large quantities of land are needed to grow the crops for biofuels and it could be needed to grow crops for food. ✓

MARKER COMMENTARY

(a) (i)The correct products and balancing each gain the mark. 2 marks.

(ii) Jessica recalls the correct conditions for the fermentation of glucose. 3 marks

(b) Jessica recalls the process name correctly and gives a well-worded answer to the next part that mentions both the hydration and the fermentation processes. 2 marks.

(c) (i) Correct definition. 1 mark.

(ii) Correct disadvantage of biofuel. 1 mark.

Note that her wording to these answers is different from Ruairi's. As long as the meaning is the same, definitions will be marked correct and sometimes there is more than one correct answer, e.g. there is more than one valid disadvantage of biofuels.

Overall Jessica gains 9 out of 9 marks. Very good recall and very good question technique.

Q&A 2 Most alcohols are readily oxidised.

(a) (i) Give the reagent(s) required to convert methanol to methanoic acid. [1]

   (ii) Draw and label a diagram of the apparatus you would use to ensure complete conversion of methanol to methanoic acid. [3]

Carboxylic acids can be converted to esters.

(b) (i) Write an equation for the reaction of methanol with methanoic acid. [2]

   (ii) State the technique you would use to separate the ester and explain, in detail, why this separation works. [3]

Interpretation

Knowledge of the reagents required to convert one organic chemical to another is essential. There are many conversions covered at AS level and so learning the reagents for these conversions is important. It is also necessary that you can recall what sort of apparatus is used for the different preparations and for the separations of the different organic products. Basically, you need to learn how to draw reflux and distillation diagrams and be able to explain how they work.

Ruairi's answer

(a) (i) Acidified potassium dichromate. ✓

   (ii)

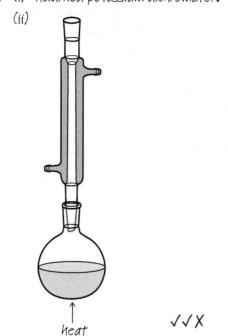

heat   ✓✓X

MARKER COMMENTARY

(a) (i) Ruairi chooses the correct combination of reagents and doesn't forget that acid is required. 1 mark.

(ii) Ruairi realises that refluxing is necessary to completely oxidise the alcohol to the carboxylic acid, he draws the apparatus correctly but forgets to label his diagram – a common mistake. Only 2 marks.

(b)(i) Ruairi gets the alcohol and acid structures correct and shows he knows the functional group for an ester (structure is clear in the equation) but forgets to add water to balance the equation. Only 1 mark

(ii) Ruairi chooses the correct separation technique and explains that this works because the chemicals have different boiling temperatures. This is not a detailed enough explanation and so he only gains 2 of the 3 marks.

Overall Ruairi gains 6 out of 9 marks. Apart from not balancing the ester equation, there is nothing wrong with Ruairi's work. He has lost marks by omission – not labelling and not explaining in detail. With more attention to detail he can greatly improve his answering technique.

(b) (i) HCOOH + CH₃OH → HCOOCH₃ ✓X

   (ii) Distillation can be used to separate the ester. This is because esters have lower boiling points than alcohols and also acids. ✓✓X

Jessica's answer

(a) (i) Acidified potassium dichromate. ✓

(ii)

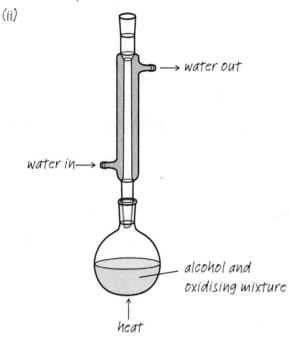

water out →

water in →

alcohol and
oxidising mixture

↑
heat

✓✓✓

(b) (i) $HCOOH + CH_3OH \longrightarrow HCOOCH_3 + H_2O$ ✓✓

(ii) Distillation can be used to separate the ester because
it has a lower boiling temperature than both methanol
and methanoic acid. Methanol and methanoic acid both
have hydrogen bonding between their molecules which
is stronger than the Van der Waals forces between
the ester molecules. And so less energy is needed to
separate the ester molecules. ✓✓✓

MARKER COMMENTARY

(a) (i) Jessica chooses the correct
combination of reagents and doesn't
forget that acid is required. 1 mark.

(ii) Jessica realises that refluxing is
necessary to completely oxidise the
alcohol to the carboxylic acid. She draws
the apparatus correctly and also labels
it, as directed by the question. 3 marks.

(b)(i) Jessica gets the alcohol and acid
structures correct and shows she
knows the functional group for an ester
(structure is clear in her equation) and
she balances the equation. 2 marks

(ii) Jessica chooses the correct separation
technique and not only mentions the
differences in boiling temperatures but
extends her answer by explaining why
the boiling temperatures are different –
hence 3 marks.

Overall Jessica gains 9 out of 9 marks.
She has through knowledge and recall
of the oxidation of alcohols and how to
convert them to esters. She also reads the
questions carefully so she notices all the
different parts of the question and makes
sure she cover them all in her answers.

2.8: Instrumental analysis

Topic summary

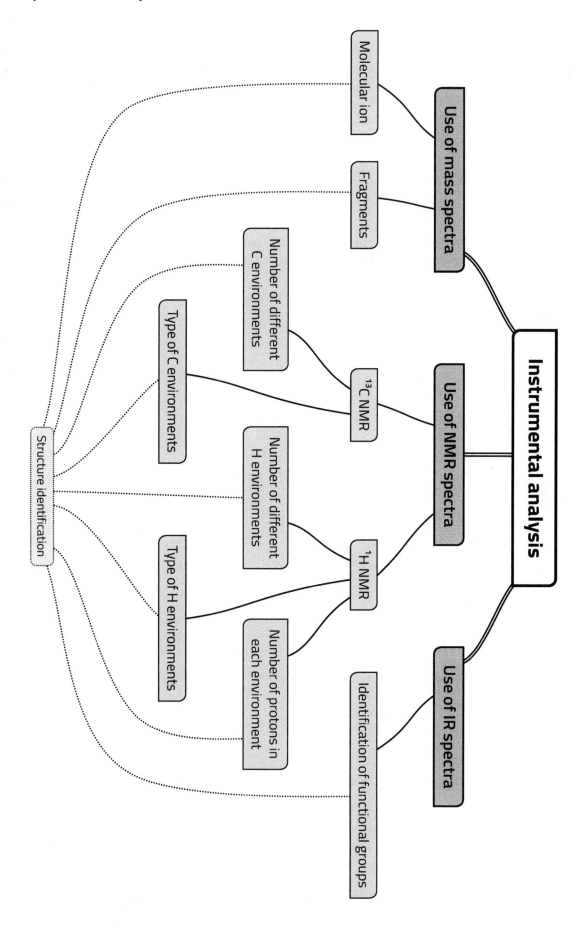

Q1 Nuclear magnetic resonance spectroscopy is a very useful tool for determining the structures of unknown organic compounds.

(a) Compare and contrast the information that can be obtained from ^{13}C and ^1H NMR spectra in order to determine the structure of an unknown compound. [5]

....

....

....

....

....

....

....

(b) Show how the information referred to in part (a) could be used to differentiate between the compounds propanone and propanal. Include diagrams of both structures as part of your answer. [6]

....

....

....

....

....

....

....

....

....

....

....

Q2 Analysis of an unknown compound X showed that X was made up of 40.7% carbon, 5.1% hydrogen and 54.2% oxygen.

(a) Use this data to determine the empirical formula of X. [2]

(b) Mass spectrometry of X gave a molecular ion peak value of m/z = 118. Determine the molecular formula of X. [2]

(c) The ^{13}C NMR spectrum of X is shown below:

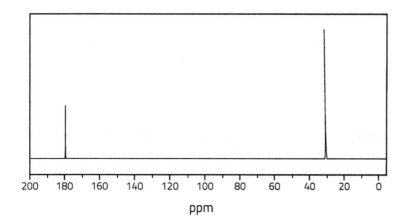

ppm

Use the number and position of the peaks to deduce the structure of X. Explain your reasoning in detail. [4]

Unit 2 Practice questions

Q3 Mass spectra can be used to determine the molecular mass of organic compounds and can also yield useful information about the structure of the compound.

(a) State how the molecular mass is determined. [1]

....

(b) What other information can be gained from mass spectra of organic compounds? [2]

....

....

(c) An unknown compound, with the molecular formula $C_5H_{10}O$, was investigated. Results showed the compound to be either pentan-2-one or pentan-3-one. Mass spectrometry was able to conclusively identify the compound as pentan-2-one. Describe two similarities and two differences between the mass spectra of the two compounds. Include the structures of the two ketones in your answer. [5]

....

....

....

....

....

....

....

....

Q4 Analysis of an unknown compound Y showed that it was made up of 24.7% carbon, 2.1% hydrogen and chlorine by mass.

(a) Use this data to determine the empirical formula of Y. [3]

(b) The mass spectrum of Y showed a group of peaks, each 2 m/z units apart, at the far-right-hand side. The peak furthest to the right had a value of m/z = 150.

(i) Explain, in detail, the significance of the group of peaks each 2 m/z units apart. [2]

....

....

(ii)   Determine the molecular formula of Y. [2]

(c)   Given that there is only one signal in the ^{1}H NMR and that Y does not decolourise bromine water, suggest a structure for Y. Explain your reasoning fully. [3]

....

....

....

....

Q5   Propene-1,3-diol, methyl ethanoate and propanoic acid all share the same molecular formula.

(a)   Draw the structures of each compound. Label each structure. [3]

(b)   Examine the following IR spectra and identify which spectra corresponds to which compound. Explain your answer. [6]

A

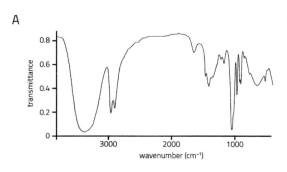

B

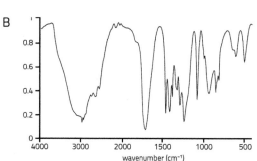

C

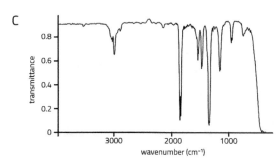

....

....

....

....

....

Question and mock answer analysis

Q&A 1 There are several straight chain isomers with the molecular formula $C_4H_8O_2$.

(a) One isomer is methyl propanoate. Give the names and structures of three other isomers that are either esters or carboxylic acids. [6]

(b) Describe the ^{13}C NMR spectra of these three isomers, explaining in detail to what extent the differences between the spectra could be used to identify each structure. [5]

Interpretation

Being able to name and draw structures are skills that cover all the organic questions in the Unit 2 paper. Being able to recognise which functional groups are isomers of each other, e.g. esters and acids, is also a common question type. Make sure you regularly practise naming and drawing isomers.

The data book is a rich source of information and will be needed to answer questions on the content covered in Section 2.8. Make sure you are familiar with all the values and units (you do not need to learn them). Many students get confused with the descriptions of the type of carbons in the ^{13}C NMR signals – make sure you know what each refers to and how to use this information in an answer.

Ruairi's answer

(a)

Butanoic acid ✓✓

Ethyl ethanoate ✓✓

Methyl propanoate ✓✗

(b) Butanoic acid would have 4 signals.

Two would be 5 to 40 ppm

There would be another 20 to 50 ppm

And the acid carbon would be about 170 ppm. ✓

Ethyl ethanoate would have 4 signals.

The C=O would be about 200 ppm.

There would then be 2 signals between 5 and 50 ppm and the carbon next to the oxygen would be about 70 ppm. ✗

Methyl propanoate would have 4 signals.

The C=O would be about 200 ppm. Then there would be 3 signals between 5 and 90 ppm. ✗

It would not be easy to identify the compounds from these spectra because they all have 4 signals and there are too many similar signals, e.g. methyl propanoate and ethyl ethanoate both have signals about 200 ppm. ✗✗

MARKER COMMENTARY

(a) Ruairi has drawn his structures using different chemical representations but the structures are unambiguous and correct, so he gains these three marks. He does not name the last ester correctly (muddling the alkyl and the acid) so he only gets 2 of the 3 naming marks.

(b) Ruairi makes a good effort but structures his answer so poorly that he only gets 1 mark – probably because he recognises that there are 4 signals in each case. Apart from a very confusing answering style, he does not link the signal values to the specific parts of the structure – it is, however, okay to give single values within ranges instead of the range itself. It is not sufficient just to quote values from the data book – they must be linked to the molecule being discussed. So sadly, he only gains 1 mark out of 5 for this whole section.

Overall Ruairi gains 6 out of 11 marks. Ruairi needs to learn how to name esters. More importantly he needs to learn how to describe the NMR signals for structures. Reading mark schemes to see what a model answer looks like is a very good idea.

Jessica's answer

(a)

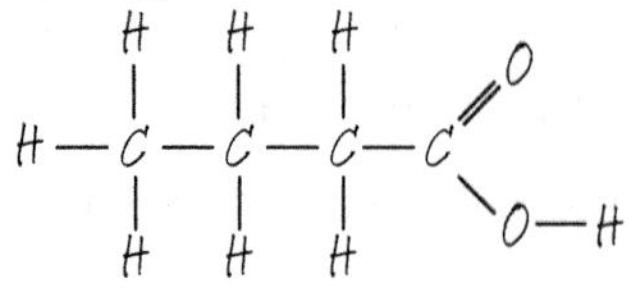

Butanoic acid ✓✓

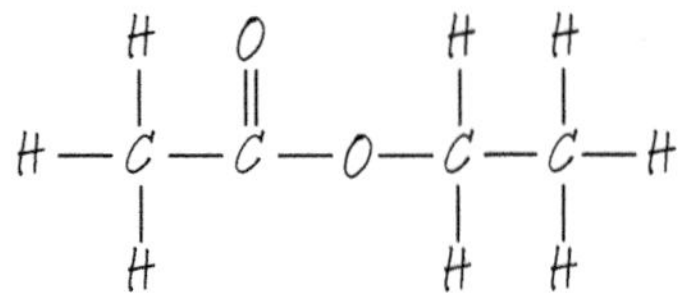

Ethyl ethanoate ✓✓

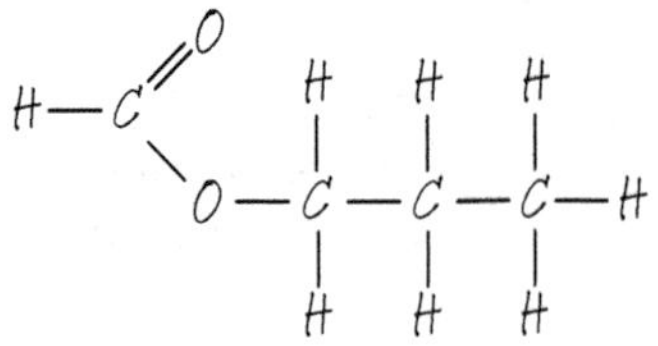

Propyl methanoate ✓✓

(b) Butanoic acid

Left to right: CH_3 5 to 40 ppm

CH_2 5 to 40 ppm (two signals)

C=O acid 160 to 185 ppm ✓

Ethyl ethanoate

Left to right: CH_3CO 20 to 50 ppm

C=O ketone 190 to 220 ppm

OCH_2 50 to 90 ppm

CH_3 5 to 40 ppm ✓

Propyl methanoate

Left to right: C=O aldehyde 190 to 220 ppm

OCH_2 50 to 90 ppm ✓

CH_2 and CH_3 5 to 40 ppm (two signals)

As you can see from this information, butanoic acid would be
identified because it did not have a signal 50 to 90 ppm and it
did not have a signal above 190 ppm. Propyl methanoate would be
identified because it had a 50 to 90 ppm signal and two signals 5
to 40 ppm, whereas ethyl ethanoate would only have one signal 5
to 40 ppm. ✓✓

MARKER COMMENTARY

(a) Jessica has drawn three structures
clearly and correctly and she has
also named them correctly. 6
marks

(b) Jessica lays out her answer well
and makes it clear what signals
would correspond to each part
of the structure. She quotes the
ranges from the data book and
links them to specific parts of
the molecular structures. She
is then able to quickly compare
the differences in the positions
of these signals and shows how
each spectrum can be assigned
to each molecule. She does this
correctly and gains all 5 marks.

Overall Jessica gains 11 out of 11.
She has good naming and drawing
skills and she has learned the
correct technique for linking data
book values to specific parts of the
structure being discussed. Another
way to do this would be to label a
diagram of the structure – probably
quicker and easier. Something to
consider next time.

Q&A 2 Analysis of an unknown organic compound X showed it contained 62.0% carbon, 10.4% hydrogen and 27.6% oxygen. X has an M_r of 58.

(a) Determine the molecular formula of X, showing all your working out. [3]

(b) The structure of X was narrowed down to be either an aldehyde or a primary alcohol. Draw and name two possible structures. [4]

(c) X was found to be an aldehyde – describe, in detail, the ^{1}H NMR spectrum of X. [4]

Interpretation

Questions involving instrumental analysis commonly require the determination of the empirical and molecular formulae. They also often involve data from chemical tests which allow you to determine what sorts of functional groups are present. Make sure you practise calculations like these and also learn all the key identification tests for alkenes, alcohols, aldehydes/ketones and carboxylic acids.

Unit 2 Practice questions

Ruairi's answer

(a) C: $58/12 \times 0.62 = 3$

H: $58/1.01 \times 0.104 = 6$

O: $58/16 \times 0.276 = 1$ ✓✓

Molecular formula $= C_3H_6O$ ✓

(b) The aldehyde will be propanal ✓

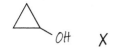

✓

The alcohol is ✗

OH ✗

(c) There would be an aldehyde signal about 9.8 ppm. ✓

There would be a signal for the CH_2 next to the aldehyde at about 2.5 ppm. ✓

There would be a signal at about 0.5 ppm for the methyl group. ✓✗

MARKER COMMENTARY

(a) Ruairi correctly demonstrates one of the ways in which molecular formula can be calculated. 3 marks.

(b) Ruairi names and draws the aldehyde correctly. His alcohol is incorrect because it is not a primary alcohol. He may have gained one mark if he had given a correct name (cyclopropanol – this isn't a required naming skill but would have been worth a guess). Only 2 marks.

(c) Ruairi's answer mentions three peaks and their positions. Whilst the wording of his answer is a little clumsy, the meaning is clear and so he gets 3 marks for this. It is acceptable to give a value within the range. However, he forgets to mention the ratio of H atoms in each environment and loses the fourth mark.

Overall Ruairi gains 8 out of 11 marks. He is good at calculations. He would have gained more marks for part (b) if he had read the question more carefully. Part (c) was reasonable but he needs to ensure he covers all aspects of ^{1}H NMR and he could do with improving the structure of answers like this. He could draw the structure and label the chemical shift for each H environment on the diagram if knowing how to word an answer is a problem.

Jessica's answer

(a)

C	H	O
62/12	10.4/1.01	27.6/16
5.16	10.3	1.73
3	6	1 ✓

Empirical formula = C_3H_6O

Mass of empirical formula = 58 ✓

So molecular formula = C_3H_6O ✓

(b) Aldehyde is propanal ✓

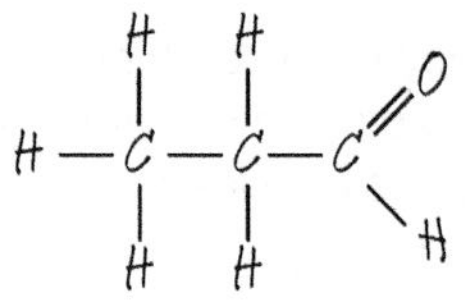

✓

Primary alcohol is prop-2-en-1-ol ✓✓

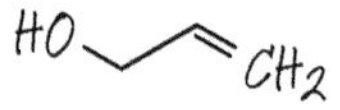

(c) There would be 3 peaks in the NMR in the ratio
3:2:1 (left to right as drawn in the structure above
for propanal) ✓

$-CH_3$ signal would be 0.1 to 2.0 ppm ✓

$-CH_2CO$ signal would be 2.0 to 3.0 ppm ✓

RCHO signal would be 9.8 ppm depending upon
concentration and solvent. ✓

MARKER COMMENTARY

(a) Jessica is competent at these calculations and
determines the correct molecular formula. 3 marks.

(b) Jessica draws and names the aldehyde correctly.
There are two possible correct alcohol structures
and Jessica gives one of them. The name is quite a
challenge but is correct. It is likely that marks would
also be awarded for prop-2-ene-1-ol as it is so
close. The alternative correct structure would be for
prop-1-en-1-ol. 4 marks.

(c) Jessica remembers to mention the 3 aspects of
^{1}H NMR she has learned about – number of peaks,
ratio of peak heights, and position of peaks. She
does so correctly – 3 marks.

Overall Jessica gains 11 out of 11. She is good at
calculations and has practised drawing and naming
structures. She has learned ^{1}H NMR thoroughly and
knows how to word her answers.

UNIT 1 PRACTICE PAPER

SECTION A

1. Sodium forms Na^+ ions in most of its compounds.

   (a) Write the electronic structure of a sodium ion. [1]

   ....

   (b) Write the equation that represents the first ionisation energy of sodium. [1]

   ....

2. Define pH. [1]

   ....

3. Magnesium and aluminium are both electrical conductors. Explain why aluminium is a better conductor than magnesium. [2]

   ....

   ....

   ....

   ....

4. Use a dot-and-cross diagram to show the bonding in calcium fluoride. [2]

5. The first line of the Balmer series of the emission spectrum of hydrogen has a frequency of 4.57×10^{14} Hz.

   (a) State which energy levels are involved in the transition that produces this line. [1]

   ....

   (b) Calculate the energy of the transition that produces this line, in $kJ\,mol^{-1}$. [2]

   Energy = ... $kJ\,mol^{-1}$

SECTION B

6. Ammonia, NH_3, phosphine, PH_3 and arsine, AsH_3, are all gaseous compounds of group 5 elements.

(a) State and explain which of these compounds will have the highest solubility. [2]

....

....

....

(b) State and explain which of these compounds will have the lowest boiling temperature. [3]

....

....

....

....

(c) Ammonia can react with H^+ ions to form ammonium ions, NH_4^+. These ions contain both covalent and coordinate bonds.

State what is meant by a coordinate bond and explain how it is different from a covalent bond. [2]

....

....

....

....

(d) Phosphine can be produced by the decomposition of phosphorous acid, H_3PO_3:

$$4H_3PO_3 \longrightarrow PH_3 + 3H_3PO_4$$
$$M_r = 82.03 \qquad M_r = 34.03 \qquad M_r = 98.07$$

Decomposition of 41.3 g of phosphorous acid produced 84.2% yield of phosphine. Calculate the mass of phosphine produced. [2]

Mass of phosphine = ... g

7. Rubidium is a group 1 element, located between potassium and caesium.

(a) Rubidium consists of two isotopes, ^{85}Rb and ^{87}Rb, with the lighter isotope making up 72.17% of naturally occurring rubidium.

Calculate the relative atomic mass of rubidium to four significant figures.
You must show your working. [2]

A_r *(Rb) =* ....

(b) Rubidium chloride has a structure that is the same as that of NaCl.

(i) Draw a labelled diagram of the structure of RbCl. [2]

(ii) Give a reason why this structure is different from that of CsCl. [1]

....

....

(c) The first ionisation energy of rubidium is lower than that of potassium and strontium.

(i) Explain why the first ionisation energy of rubidium is lower than that of strontium. [2]

....

....

....

....

(ii) Explain why the first ionisation energy of rubidium is lower than that of potassium. [2]

....

....

....

....

(d) A synthetic isotope of rubidium, ^{84}Rb, can decay in several different ways.

(i) The four different decay routes are shown below. Complete the diagram to show the mass numbers and symbols of the isotopes formed. [2]

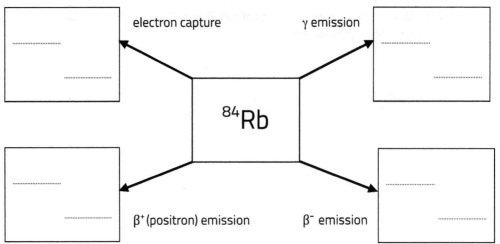

(ii) Another type of radiation is α-emission. State what is meant by an α-particle. [1]

....

....

8. Nitrosyl chloride, NOCl, is a molecule that can be formed in several ways.

(a) One method of preparing nitrosyl chloride is to react nitric oxide (NO) with chlorine. This is an exothermic reaction:

$$2NO + Cl_2 \rightleftharpoons 2NOCl$$

(i) Write an expression for the equilibrium constant, K_c, for this reaction. [1]

(ii) At 35 °C the numerical value of the equilibrium constant is 65000. The concentration of NO is 0.015 mol dm^{-3} and Cl_2 is 0.012 mol dm^{-3}.

Calculate the concentration of nitrosyl chloride. [2]

Concentration of NOCl = ... mol dm^{-3}

(iii) The experiment is repeated at a higher temperature. State and explain the effect of this on the value of K_c. [2]

....

....

....

....

(b) Nitrosyl chloride has nitrogen bonded to two different atoms as shown below:

$$O=N-Cl$$

A student states that this molecule should be linear as the nitrogen is bonded to two other atoms only. Is the student correct? Justify your answer. [3]

....

....

....

....

(c) The electronegativity values of the atoms present in nitrosyl chloride are given:

Element	N	O	Cl
Electronegativity value	3.0	3.5	3.0

The bonds between N and O are classed as polar covalent, whilst the N–Cl bond is non-polar covalent.

Explain why these atoms form covalent bonds and why some of these bonds are polar and some are not. [2]

....

....

....

(d) Nitrosyl chloride reacts with cold water to form an acidic solution:

$$NOCl + H_2O \rightarrow HCl + HNO_2$$

HCl is a strong acid and HNO_2 is a weak acid. Explain the difference between strong and weak acids. [2]

....

....

....

9. A mixture of sodium carbonate, Na_2CO_3, and sodium hydrogencarbonate, $NaHCO_3$, can be analysed by a titration against hydrochloric acid using two indicators.

STEP 1: The first indicator changes colour when the carbonate ions have reacted with H^+ ions to form hydrogencarbonate ions. This is endpoint 1.

$$H^+ + CO_3^{2-} \rightarrow HCO_3^-$$

STEP 2: The second indicator changes colour when all the hydrogencarbonate ions have reacted with H^+ ions. This include the hydrogencarbonate ions that were present at the start as well as the hydrogencarbonate ions formed from carbonate ions in the first step. This is endpoint 2.

$$H^+ + HCO_3^- \rightarrow H_2O + CO_2$$

Ben performs an experiment where 25.0 cm³ samples of the solution containing Na_2CO_3 and $NaHCO_3$ are titrated against hydrochloric acid of concentration 0.120 mol dm⁻³. The results of the experiment are given in the table:

	1	2	3	4
Initial burette reading / cm³	0.00	0.40	1.20	0.25
Burette reading at endpoint 1 / cm³	15.50	15.85	16.75	15.75
Burette reading at endpoint 2 / cm³	42.85	42.75	43.70	42.75
Volume of acid for reaction in step 1 / cm³				
Volume of acid for reaction in step 2 / cm³				

(a)   Complete the table and calculate the mean volumes required for reaction in step 1 and step 2.   [2]

Mean volume in step 1 = ... cm³

Mean volume in step 2 = ... cm³

(b)   State, giving a reason, which of the two volumes in part (a) you consider to be the most accurate.   [1]

....

....

(c)   Alice suggests using a lower concentration of hydrochloric acid in the experiment and Carwyn suggests using a higher concentration. Their teacher tells them that they should not make these changes.

(i)   Give one reason that they should not follow Alice's suggestion.   [1]

....

....

(ii)   Give one reason that they should not follow Carwyn's suggestion.   [1]

....

....

(d)   (i)   Use the mean volume from step 1 to calculate the concentration of Na_2CO_3 in the solution mixture.   [3]

Concentration of Na_2CO_3 = ... mol dm⁻³

(ii) Use the results and your answer to part (d)(i) to find the concentration of $NaHCO_3$ in the original solution. [3]

Concentration of $NaHCO_3$ = ... mol dm^{-3}

10. (a) A student is provided with a selection of four unlabelled white solid samples. They know that the samples are four out of the following compounds.

Calcium carbonate   Sodium iodide   Magnesium carbonate

Calcium chloride   Sodium bromide   Magnesium hydroxide

Outline a plan to identify the compounds as efficiently as possible. [QER 6]

....

....

....

....

....

....

....

....

....

....

....

(b) Both magnesium carbonate and calcium carbonate can undergo thermal decomposition when heated. The temperature needed for decomposition of calcium carbonate is 850 °C.

Suggest a suitable temperature for the decomposition of magnesium carbonate. Give a reason for your answer. [1]

....

....

....

(c) If chlorine gas is bubbled through a solution of sodium bromide a displacement reaction occurs to produce sodium chloride.

(i) Write an ionic equation for this reaction. [1]

....

(ii) Explain why chlorine gas will react with sodium bromide whilst iodine will not. [2]

....

....

....

11. Georgeite is a rare mineral found in a disused copper mine in Snowdonia. It has a formula of $Cu_a(CO_3)_b(OH)_c.dH_2O$.

In an experiment a test tube containing 0.0140 mol of georgeite was heated gently until it reached constant mass. This caused a decrease in mass as the water of crystallisation was lost. Further heating at a higher temperature caused the carbonate and hydroxide to decompose in a manner similar to the thermal decomposition of calcium carbonate and calcium hydroxide. This left a sample of pure dry CuO. The gases produced during the reaction are passed through a tube cooled to 0 °C to condense the water vapour and the volume of carbon dioxide gas was measured at 0 °C and 1 atm pressure.

These are the results of the experiment:

Mass of empty test tube   = 38.724 g
Mass of test tube + georgeite at start of experiment   = 43.332 g
Mass of test tube + georgeite after gentle heating   = 41.818 g
Mass of test tube + copper oxide remaining after sustained heating = 40.950 g
Volume of carbon dioxide gas produced   = 314 cm³

(a) Calculate the mass of georgeite present at the start of the experiment. [1]

Mass of georgeite = ... g

(b) Calculate the percentage error in your measurement in part (a).
You must show your working. [1]

Percentage error in mass = ... %

(c) Calculate the number of molecules of water present in the formula. [2]

d = ...

(d) Calculate the number of moles of carbon dioxide produced and hence the value of b. [3]

Moles of CO_2 = ... mol

c = ...

(e) Find the value of a. [3]

a = ...

(f) Find the M_r of the initial georgeite and use this value, in addition to your previous calculations, to give the formula of georgeite.
You must show your working. [3]

M_r of georgeite = ...

Formula of georgeite = ...

(g) Show that the oxidation state of copper in this compound is +2. [2]

....

....

....

....

....

UNIT 2 PRACTICE PAPER

(80 Marks)

SECTION A

1. Draw the displayed structure of a primary alcohol with 4 carbon atoms.   [1]

2. Paracetamol has the structure shown below. What is its molecular formula?   [1]

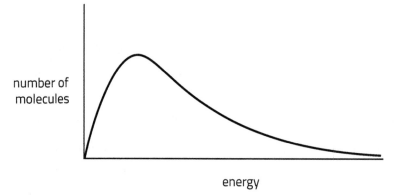

....

3. Catalysts are used to speed up reactions. Annotate the Boltzmann distribution below to help explain how a catalyst works.   [2]

number of molecules

energy

....

....

....

4. Ethanol can be converted into ethanoic acid under the correct practical conditions.

    (a) Name the type of reaction.   [1]

....

(b) Ethanoic acid reacts with sodium carbonate in a 2:1 ratio. Calculate the volume of 0.5 mol dm^{-3} solution of sodium carbonate that would neutralise 14.0 cm^3 of 1.0 mol dm^{-3} ethanoic acid. [1]

(c) Write an equation for the reaction between ethanoic acid and propan-2-ol showing clearly the structure of the organic product. [1]

....

(d) Calculate the atom economy for making this organic product. [1]

....

....

5. Give the full name of the following compound: [2]

....

SECTION B

6. A student carried out a simple laboratory experiment to measure the enthalpy of combustion of ethanol. The student found that the temperature of 100 g of water increased by 16.0 °C when 0.46 g of ethanol was burned in air and the heat produced was used to warm the water.

(a) Draw a labelled diagram of the apparatus that the student would use. [3]

(b) Use the results to calculate the value obtained by the student for this enthalpy change. Give your answer in kJ mol^{-1}. [3]

(c) The table below shows the data book values for the enthalpy of combustion of several primary alcohols:

Alcohol	Enthalpy of combustion / kJ mol^{-1}
Methanol	−726
Ethanol	−1367
Propan-1-ol	−2017
Butan-1-ol	

(i) Give one reason, other than heat loss, why the value obtained from the student's results is less exothermic than the data book value. [1]

(ii) Use the values in the table to predict a value for butan-1-ol. [1]

(iii) In terms of bonds broken and bonds formed, explain why the values in the table follow a predictable pattern. [2]

7. The kinetics of the oxidation of iodide ions by hydrogen peroxide in acid solution can be studied using an 'Iodine clock' reaction:

$$H_2O_2 + 2I^- + 2H^+ \rightarrow I_2 + 2H_2O$$

In this experiment the concentration of hydrogen peroxide was varied with all other concentrations and the temperature being kept constant. The rate of reaction for each experiment was determined by timing until the distinctive blue-black colour appeared. The results are shown in the table:

Concentration of hydrogen peroxide / mol dm^{-3}	Time for appearance of blue-black colour / s	Rate / s^{-1}
0.010	32.5	
0.008	39.1	
0.006	55.0	
0.004	85.9	
0.002	129.0	

(a) Use rate = 1/time to calculate the rate for each experiment and complete the table.   [1]

(b) On the axes below, plot the concentration of hydrogen peroxide against rate and draw a suitable line.   [3]

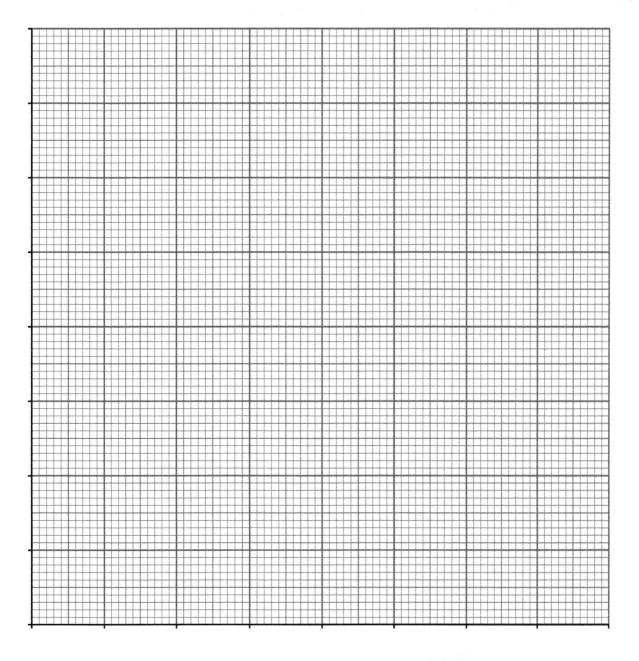

(c) From the graph deduce the relationship between concentration of hydrogen peroxide and the rate of reaction. [1]

....

....

(d) Explain, in terms of collision theory, the effect of increasing the concentration of hydrogen peroxide on the rate of reaction. [2]

....

....

....

8. This question is about ethanol.

(a) One example of a process by which ethanol can be made is shown below:

$$CH_3CH_2Br \rightarrow CH_3CH_2OH$$

(i) Name and outline the mechanism for this reaction.

Name of mechanism .... [1]

Mechanism: [2]

(ii) Suggest a reason, other than safety, why this method is not used in industry to make ethanol. [1]

....

....

(iii) It is also possible to carry out the reverse process as represented by the reaction scheme below:

$$CH_3CH_2OH \xrightarrow{\text{Reaction 1}} \text{Substance X} \xrightarrow{\text{Reaction 2}} CH_3CH_2Br$$

Name the mechanism and give a suitable reagent for both reactions.

Reaction 1:

Name of mechanism .... [1]

Reagent .... [1]

Reaction 2:

Name of mechanism .... [1]

Reagent .... [1]

(b) The following reaction is used to make ethanol in industry:

$$H_2C=CH_2 \rightarrow CH_3CH_2OH$$

(i) Name this type of reaction. [1]

....

(ii) Give the name of a catalyst for this reaction and two other conditions necessary to produce ethanol in a high yield. [3]

....

....

....

9.   (a) The reaction between ethane and chlorine is similar to the reaction of chlorine with methane.

(i) Name the type of mechanism and give the essential condition for the reaction to occur. [2]

Type of mechanism ....

Essential condition ....

(ii) Write the equations for the following steps in the mechanism for the reaction between chlorine and ethane. [4]

Initiation step

....

First propagation step

....

Second propagation step

....

A termination step leading to the formation of 1,2-dichloroethane.

....

(b) Chloroethane is a colourless gas whereas 1,2-dichloroethane is a colourless liquid.

(i) Explain the difference in boiling temperatures of these two halogenoalkanes. [3]

....

....

....

(ii) Calculate the volume occupied by 3.75 g of chloroethane at 100 °C and 2 atmospheres. Give your answer in cm³. [4]

(iii) What volume of 1,2-dichlorethane liquid would contain the same number of molecules as 3.75 g chloroethane gas? The density of 1,2-dichloroethane is 1.25 g/cm³. [2]

....

10.   2-bromobutane can be prepared according to the following equation:

$$CH_3CH_2CH{=}CH_2 + HBr \rightarrow CH_3CH_2CHBrCH_3$$

The enthalpy change for this reaction can be calculated in a variety of ways.

(a) State Hess's law.   [1]

....

....

(b) Draw an energy cycle (using the equation below) to show how enthalpy of formation values could be used to determine the enthalpy change for the reaction to make 2-bromobutane. Label each arrow showing what values would be used and include any relevant state symbols.   [3]

$$CH_3CH_2CH{=}CH_2 + HBr \rightarrow CH_3CH_2CHBrCH_3$$

(c) Bond enthalpy values can also be used to determine the enthalpy change for the reaction to make 2-bromobutane.

(i)   State what is meant by average bond enthalpy.   [2]

....

....

....

(ii) Use the values in the table below to calculate the enthalpy change for the reaction.   [3]

$$CH_3CH_2CH{=}CH_2 + HBr \rightarrow CH_3CH_2CHBrCH_3$$

Bond	Average bond enthalpy / kJ mol^{-1}
C–H	412
C–C	348
C=C	612
H–Br	366
C–Br	276

....

(d) Explain why the values calculated by the methods in (b) and (c) would be different. [1]

....

....

....

11. Isomers are common in organic chemistry.

(a) Give the meaning of the term *structural isomers*. [1]

....

....

(b) The following compounds can be distinguished by carrying out test tube reactions. In each case, suggest a suitable reagent or reagents that could be added to each separately in order to distinguish them. Give observations for each compound with the reagent(s) chosen.

(i) [3]

$$H_3C - \overset{\overset{\displaystyle CH_3}{|}}{\underset{\underset{\displaystyle CH_3}{|}}{C}} - \overset{\overset{\displaystyle H}{|}}{\underset{\underset{\displaystyle H}{|}}{C}} - \overset{\overset{\displaystyle OH}{|}}{\underset{\underset{\displaystyle H}{|}}{C}} - H$$

M

$$H_3C - \overset{\overset{\displaystyle OH}{|}}{\underset{\underset{\displaystyle CH_3}{|}}{C}} - \overset{\overset{\displaystyle H}{|}}{\underset{\underset{\displaystyle H}{|}}{C}} - \overset{\overset{\displaystyle CH_3}{|}}{\underset{\underset{\displaystyle H}{|}}{C}} - H$$

N

Reagent(s) ....

Observations...

....

....

....

(ii) [3]

$$\underset{O}{\overset{CH_3O}{\diagdown}} C - CH_3$$

P

$$\underset{O}{\overset{HO}{\diagdown}} C - CH_2CH_3$$

Q

Reagent(s) ....

Observations...

....

....

....

(iii) [3]

R   S

Reagent(s) ....

Observations...

....

....

....

(c) Compounds R and S were also studied using IR and ^{1}H NMR spectroscopy.

(i) Describe the differences you would expect to see in the IR spectra. [2]

....

....

....

(ii) Describe the differences you would expect to see in the ^{1}H NMR spectra, referring to numbers of peaks and the areas under the peaks. [4]

....

....

....

....

....

Answers

Practice questions: Unit 1: The Language of Chemistry, Structure of Matter and Simple Reactions

Section 1.1: Formulae and equations

Q1 (a) (i) Ba^{2+} (aq) + SO_4^{2-} (aq) → $BaSO_4$ (s) 1 mark for equation. 1 mark for state symbols.

(ii) Fe^{2+} (aq) + $2OH^-$ (aq) → $Fe(OH)_2$ (s) 1 mark for equation. 1 mark for state symbols.

(b) (i) Cl_2 (g) + $2Br^-$ (aq) → $2Cl^-$ (aq) + Br_2 (aq) 1 mark for equation. 1 mark for state symbols.

(ii) Oxidation state of chlorine goes from 0 at start to −1 at end. This is reduction as oxidation state becomes more negative (1 mark).

Oxidation state of bromine goes from −1 at start to 0 at end. This is oxidation as oxidation state becomes more positive (1 mark).

Q2 (a) $CuCO_3$ + $2CH_3COOH$ → $Cu(CH_3COO)_2$ + H_2O + CO_2 (1 mark for correct formulae, 1 for balancing).

(b) $S_2O_3^{2-}$ = +2; S = 0; SO_2 = +4 (1 mark for 2 correct, 2 for all 3 correct).

Q3 (a) Al_2O_3 + $3H_2SO_4$ → $Al_2(SO_4)_3$ + $3H_2O$ (1 mark for correct formulae, 1 for balancing).

(b) (i) H^+ (aq) + OH^- (aq) → H_2O (l).

(ii) Al^{3+} (aq) + $3OH^-$ (aq) → $Al(OH)_3$ (s).

Q4 (a) $6PbO$ + O_2 → $2Pb_3O_4$ (1 mark for correct formulae, 1 mark for balancing).

(b) Oxidation state of lead at the start = +2, at end = +2.67 (1 mark, accept fractions or any rounded number with at least one decimal place) (1).

Oxidation state of lead has become more positive so has been oxidised (1).

Section 1.2: Basic ideas about atoms

Q1 Nitrogen or any other group 5 element.

Q2 $1s^2 \, 2s^2 \, 2p^6$

Q3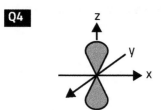

Q4

Q5 Group 4 (1) as the most significant increase is between the 4th and 5th ionisation (1).

Q6 (a) Both processes involve electrons which have very little mass / Mass number only affected by loss or gain of protons or neutrons.

(b) As, Kr

Q7 233, Am (1); 229, Pu (1).

Q8 (a) (2 for all three columns correct, 1 for two columns correct).

	α	β	γ	positron
Change in atomic number	−2	+1	0	−1
Change in mass number	−4	0	0	0

(b) β deflected towards positive plate, γ not deflected, positron deflected towards negative plate similar amount to β. (2 for all three correct, 1 for two correct.)

(c) α particles are blocked by glass bottle so are contained (1) but once in the body are very ionising and destroy/damage cells/biological molecules (1).

(d) Tracers to follow biological molecules around the body (e.g. iodine for thyroid function) / radiotherapy to kill cancerous cells (e.g. cobalt-60).

Q9 (a) 1 mark for correct protons and electrons, 1 mark for correct neutrons.

	^{234}U	^{235}U	^{238}U
Number of protons	92	92	92
Number of neutrons	142	143	146
Number of electrons	92	92	92

(b) (i) Th, 230, 90

(ii) 4 half-lives (1), 1/16th remains (1), 3.125×10^{-4}% (1).

Q10 (a) α, β⁻, β⁻

(b) The half-lives for decay of ^{228}Ra and ^{228}Ac are much shorter than that of ^{232}Th (1). ^{228}Ra and ^{228}Ac decay as fast as they are produced so do not build up (1).

Q11 (a) (i) He nucleus / He^{2+}

(ii) Causes mutations / cancers / ionises DNA.

(iii) Incorrect (1) as gamma radiation could also be emitted as it doesn't affect the atomic or mass numbers (1).

(b) Two half-lives (1) 0.55 g (1).

Q12 (a) Removed from a different shell (1) closer to the nucleus with no shielding OR greater effective nuclear charge (1).

(b) Nitrogen has more protons / greater nuclear charge (1). Greater attraction holding electron (1).

(c) Sixth electron of oxygen is in the second shell (1) so it has greater shielding / further from the nucleus (1).

Q13 (a) Arrow pointing upwards from I to H.

(b) Arrow pointing downwards from F to H.

(c) Arrow pointing upwards from I to A.

(d) $E = 2.179 \times 10^{-18}$ J (1); $f = 3.287 \times 10^{15}$ (1); wavelength = 91.26 nm (1).

Q14 (a) Ne, Mg

(b)

↑↓		↑↓		↑↓	↑↓	↑↓		

$1s^2$ $2s^2$ $2p^6$ $3s^0$

(c) Sodium (1). Outer electron in potassium is further from the nucleus and shielding is more which overcomes greater nuclear charge (1).

(d) Sodium, magnesium, neon (1). Neon is highest as electron removed from different shell, closer to nucleus with much less shielding (1). Magnesium is greater than sodium as it has a greater nuclear charge with similar shielding and electron similar distance from nucleus (1).

Q15 (a) Li (g) → Li⁺(g) + e⁻ [Must show state symbol] (1).

(b) Helium has a greater nuclear charge than hydrogen (1) but outer electrons of both are the same distance from the nucleus / in the same shell with the same (no) shielding (1).

(c) Outer electron of nucleus is in a different shell to helium (1) further from nucleus with greater shielding which overcomes the greater nuclear charge (1).

(d) 200–480 kJ mol⁻¹ (1). Ionisation energy decreases down group so must be lower than Li (1).

(e) 1100–1370 kJ mol⁻¹ (1). Outermost electron in oxygen is paired, and in nitrogen it is unpaired (1). Electron–electron repulsion in oxygen reduces ionisation energy below the value for nitrogen (1).

Q16 (a) Pattern of gradual increases, with jumps after third and eleventh value (1).
Gradual increase due to increase in proton : electron ratio OR charge on ion (1).
Jumps due to different shells (1).
So electrons closer to nucleus with less shielding (1).

(b) Outer electron in magnesium in s sub-shell and for aluminium it is in a p sub-shell (1).
Outer electron of aluminium therefore has slightly increased shielding and lower ionisation energy (1).

(c) Electron lost in both cases is innermost electron / in same shell with no shielding (1).
But aluminium has a greater nuclear charge than magnesium so greater attraction between nucleus and electron (1).

Section 1.3: Chemical calculations

Q1 A mole is the amount of substance that contains the same number of particles (1) as there are atoms in 12 g of carbon-12 (1).

Q2 N_2O_5

Q3 Avogadro's number is the number of atoms present in 12 g of carbon-12.

Q4 (a) 0.02 mol (1); 1.20×10^{22} molecules (1).

(b) 4.82×10^{22} atoms.

Q5 The empirical formula is the simplest ratio of atoms of each element present in a compound.

Q6 49.31% ^{81}Br (1); 79.99 (1 mark for answer, 1 mark for 4 significant figures).

Q7 33.3% Mn (1); Empirical formula = Na_2MnO_4 (1).

Q8 (a) M_r (Na_2CO_3) = 106 (1); x = 10 (1).

(b) M_r of hydrate = 232 (1); x = 7 (1).

Q9 Relative molecular mass is the average mass of a molecule of a substance (1) relative to 1/12th the mass of an atom of carbon-12 (1).

Q10 (a) Division by A_r values (1); C_2H_2O (1).

(b) M_r (C_2H_2O) = 42.02 (1); Molecular formula = $C_6H_6O_3$ (1).

Q11 (a) CO_2H

(b) Moles oxalic acid = 0.0691 mol (1); moles CO_2 = 0.1382 mol (1); volume CO_2 = 3096 cm^3 (1).

Q12 (a) Moles XeO_3 = 0.151 mol (1); Moles gas = 0.377 mol (1); Volume = 8.78 dm^3 (1).

(b) Na_2XeF_8

(c) Moles XeF_2 = 0.0592 mol (1); Moles SiF_4 = 0.0296 mol (1); Mass SiF_4 = 3.08 g (1).

Q13 (a) Moles of NH_3 in 100 cm^3 = 1.820 mol (1); Concentration = 18.20 mol dm^{-3} (1).

(b) 44.6 dm^3

Q14 (a) CH_2O

(b) (i) M_r of glucose is 180 (1); 48.3% (1).

(ii) Moles glucose = 2.778×10^{-2} mol (1); Moles oxygen gas = 0.1667 mol (1); Volume = 4.25 dm^3 (1).

Q15 (a) 10 dm^3 C_2H_6 forms 20 dm^3 CO_2 and 30 dm^3 H_2O (1); 10 dm^3 reacts with 35 dm^3 O_2 (1);
Total gas = 20 + 30 + 25 = 75 dm^3 (1).

(b) (i) Volume of gas is proportional to temperature (1); Water will change from gas to liquid, reducing the amount of gas (1).

(ii) Volume of CO_2 and O_2 at 650 K = 45 dm^3 (1); Volume at 300 K = 20.77 dm^3 (2).

Q16 (a) Relative isotopic mass is the mass of one atom of an isotope relative to 1/12th of the mass of an atom of carbon-12.

(b) 1.0080 (2 for correct answer, 1 for 5 significant figures).

Q17 (a) Electron gun

(b) Key content:

- Sample is heated to vaporise.
- System is under vacuum to aid with forming vapour / prevent beam being blocked by molecules in air/ prevent molecules in air swamping detector.
- Electron gun fires fast electrons at sample to remove electrons and form positive ions.
- Negatively charged plates accelerate positive ions / focus beam.
- Electromagnet deflects ions according to mass and charge.
- Only ions with correct m/z hit the detector and can be detected.
- Changing current changes amount of deflection hence which ions are detected.

(c) (i) $118 = CH^{35}Cl_3^+$; $120 = CH^{37}Cl^{35}Cl_2^+$; $122 = CH^{35}Cl^{37}Cl_2^+$; $124 = CH^{37}Cl_3^+$

(ii) ^{35}Cl is ¾ of the atoms, ^{37}Cl is ¼ of the atoms (1).

3 chlorine atoms so proportions are $(¾)^3 : (¼)^3$ (1) = 27/64 : 1/64 giving a 27:1 ratio (1).

Q18 (a) $M_r (SOCl_2) = 119$; $M_r (CH_3COOH) = 60$; $M_r (CH_3COCl) = 78.5$ (1);
% atom economy= 43.9% (1).

(b) Moles ethanoic acid = 0.2067 mol (1); theoretical mass = 16.22 g (1); % yield = 69.0% (1).

Q19 (a) $M_r (MgSO_4) = 120.3$ (1); Solubility in g dm^{-3} = 351.2 g dm^{-3} (1); Mass in 250 cm^3 = 87.8 g (1).

(b) Moles ppt = $(4.17 - 2.92) \times 0.25 = 0.3125$ mol (1) $M_r (MgSO_4.7H_2O) = 246.44$ (1); mass = 77.0 g (1).

(c) $M_r = 246.44 \times 0.708 = 174.32$ (1). Amount of water = 72.11 (1) x = 4 (1).

Q20 (a) 359 g dm^{-3} (b) 1.726 mol dm^{-3} (c) 40.21 g.

Q21 mol NaOH = 5.067×10^{-3} mol (1); mol HX = 5.067×10^{-3} mol (1); conc HX = 0.2027 mol dm^{-3} (1); conc in g dm^{-3} = 32 g dm^{-3} (1); $M_r = 157.9$ (1).

Q22 (a) Final volume of NO = 0 dm^3, CO = 5 dm^3 (1) CO_2 = 10 dm^3, N_2 = 5 dm^3 (1) Change = −5 dm^3 (1).

(b) Time = 1440 minutes so volume = 14400 dm^3 (1); Moles of $BaSO_4 = 1.648 \times 10^{-3}$ (1); Moles $SO_2 = 1.648 \times 10^{-3}$ (1); Volume of $SO_2 = 0.461$ dm^3 (1); Percentage = 3.2×10^{-3} % (1).

Q23 (a) (i) 249.6 (ii) 2.496 g.

(b) (i) Moles Zn = 0.0130 mol and mol $CuSO_4 = 0.010$ mol (1) so Zn is in excess (1).

(ii) 0.635 g

Q24 (a) (i) 2.842 g (ii) 0.035% (iii) Greater percentage error as error in smaller masses is greater.

(b) Mass $CO_2 = 1.013$ g (1); moles $CO_2 = 0.023022$ mol (1); M_r of carbonate = 123.44 (1); Cu (1).

Section 1.4: Bonding

Q1 Van der Waals, hydrogen bonding, covalent bond.

Q2 SF_6, NCl_3, CCl_4, BCl_3, $BeCl_2$

Q3

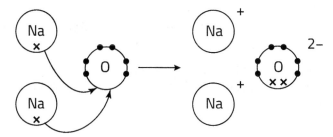

Q4 F_2, Cl_2, NH_3, NaCl

Q5 (a) The ability of an atom to attract a pair of electrons in a covalent bond.

(b) δ− O−H δ+   δ+ C=O δ−   δ+ H−F δ−   δ+ C−F δ−

(c) H−F (1) as the electronegativity difference is the largest (1).

Q6 (a)

```
 •• ××
•F× F ×
 •• ××
```

(b)

H–F (1)   F–F (1)

No electronegativity difference in F_2 so electrons are distributed evenly. Fluorine is more electronegative than hydrogen so has greater electron density (1).

(c)

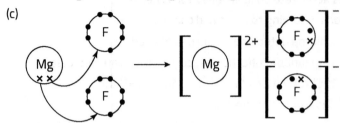

Q7 (a)

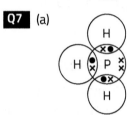

(b) Pyramidal

Q8 Down the group the molecules have more electrons (1) so the Van der Waals forces between the molecules are greater (1).

Q9 C, A, B (1); C is lowest as it does not form hydrogen bonding between molecules whilst the others do (1); Hydrogen bonding is the strongest intermolecular force (1); More Van der Waals forces between molecules of B than A so it has a higher boiling temperature (1).

Q10 (a)

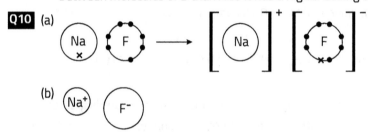

(b) (Na⁺) (F⁻)

(c) Strong forces of attraction between oppositely charged ions (1) so a lot of energy is needed to overcome these (1).

Q11 (a) H ⦂ H

(b) Repulsion between the two nuclei (1); repulsion between the two electrons (1); attraction between the two electrons and both nuclei (1).

(c) Movement of electrons causes temporary dipoles (1) and these include dipoles in adjacent molecules (1). Dipoles are small and temporary so the attraction between them is weak (1).

Q12 Electron pairs repel to be as far from each other as possible (1); Lone pairs repel more than bonded pairs (1); SF_6 has 6 bond pairs and no lone pairs (1); SF_2 has 2 bond pairs and 2 lone pairs (1); SF_6 – octahedral, SF_2 – V-shaped/bent (1).

Q13 Key points:

- Intermolecular force between δ+ on hydrogen and lone pair on atom on other.
- Strongest intermolecular forces.
- Needs hydrogen bonded to F/O/N as these are the most electronegative elements.
- Large dipole with δ– on F/O/H and δ+ on H.
- Hydrogen nucleus is exposed / low electron density around hydrogen nucleus.
- Hydrogen bonding with water molecules makes ammonia soluble.
- Hydrogen bonding between molecules leads to higher boiling temperatures.
- Can obtain many marks from diagrams.

Q14 (a) Boron atom has six valence electrons / fewer than eight outer electrons.

(b) Pair of shared electrons (1) with both electrons coming from the same atom (1).

(c) (i) 120°.

(ii) Boron goes from having three bonding pairs to four bonded pairs, and more bond pairs means a smaller angle between them.

(iii) Lone pair in NH_3 and none in CH_4 (1) and lone pairs repel more than bonded pairs (1).

Q15 (a) Boiling temperatures increase as the number of carbon atoms increases (1). All have the same amount of hydrogen bonding (1) but more carbons leads to more/stronger Van der Waals forces between molecules (1) and these require more energy to overcome them (1).

(b) 155–168 °C.

(c) Pentanoic acid, butanoic acid, propanoic acid (1). All of the acids contain −COOH which can form hydrogen bonds with water molecules (1). The carbon chain cannot form hydrogen bonds / only forms Van der Waals forces between molecules (1). As the carbon chain increases in length, a larger proportion of the molecule cannot form hydrogen bonds so solubility decreases (1).

Q16 The boiling temperature of water is significantly higher than the other hydrides, drops to H_2S then increases down the group (1); Only H_2O has hydrogen bonding between molecules (1). Hydrogen bonding is stronger than Van der Waals forces (1); Rest of molecules only have Van der Waals forces between molecules (1) and these increase in strength down the group as the number of electrons increases (1).

Section 1.5: Solid structures

Q1 Melting diamond requires breaking covalent bonds, but melting iodine requires breaking Van der Waals forces between molecules (1). Van der Waals forces are much weaker than covalent bonds (1).

Q2 Ionic compounds conduct when the ions can move and carry charge (1). The ions are in fixed positions in the solid but can move when molten (1).

Q3 (a) Similarity – Cs^+ ions arranged in a regular lattice in both (1).
Difference – sea of delocalised electrons in Cs(s), anions in regular lattice is CsCl (1).

(b) Cs conducts when solid, CsCl doesn't conduct as a solid (only as a liquid) (1).

Q4 (a) Clear arrangement of ions showing 6:6 pattern (1) Ions correctly labelled as Na^+ and Cl^- OR sodium ion and chloride ion (1).

(b) (i) Calcium chloride cannot as there are different numbers of calcium and chloride ions OR there are two chlorides for each calcium ion and NaCl structure contains equal numbers of cations and anions (1).

(ii) The melting temperature depends on the strength of the forces between the ions. Stronger forces between ions lead to higher melting temperature (1).

The charges on Ca^{2+} and O^{2-} are greater than the charges of Na^+ and Cl^- so the forces of attraction in CaO are greater (1).

(c) Water molecules with $\delta+$ on H and $\delta-$ on O surrounding Na^+ and Cl^- with H pointing at chloride ions and O pointing at sodium ion. (1 mark for Na and one for Cl).

Q5 (a) Any FOUR from:
- Both have carbon atoms joined by covalent bonds.
- Both are giant covalent structures / macromolecules.
- Graphite has delocalised electrons, diamond does not.
- In graphite each carbon bonded to three others and in diamond each carbon is bonded to four others.
- Graphite has layers of hexagons.
- Bonds arranged tetrahedrally in diamond.

(b) Must break covalent bonds to melt diamond and graphite (1). Covalent bonds are very strong and require a lot of energy to break (1).

(c) Graphite has delocalised electrons (between layers) that can move and form a current. (1)

Q6 (a)

	Ionic bonds	Covalent bonds	Coordinate bonds	Van der Waals forces	Hydrogen bonds
Iodine, I_2 (s)		✓		✓	
Water, H_2O (s)		✓		✓	✓

1 mark per correct row

(b) Need to break Van der Waals forces between molecules (1). These are the weakest intermolecular forces (1).

(c) (i) Particles in a solid are usually closer together than a liquid (1) so volume is smaller and hence density is greater (1).

(ii) In ice hydrogen bonds hold the molecules together in an open (tetrahedral) structure (1) so molecules are further apart in ice than they are in water (1). The volume of ice is greater than water, so the density is lower than water (1).

Q7 (a) Metallic bonding has a lattice of metal ions in a sea of delocalised electrons OR the forces of attraction in metals are between delocalised electrons and the metal ions (1).

There are more outer electrons in group 2 atoms than in group 1 atoms so there are more delocalised electrons in group 2 metals (1).

As there are more delocalised electrons/ions have a +2 charge for group 2 metals, the forces are stronger, so the metals are harder (1).

(b) Atoms at the end of the *d*-block have the most outer shell electrons OR copper has 11 outer shell electrons (1).

More delocalised electrons that can carry charge leads to greater conductivity (1).

Section 1.6: The periodic table

Q1 (a) $3Fe + 2Cl_2 \rightarrow 2FeCl_3$

(b) Add silver nitrate solution (1) and gives a white precipitate (1).

Q2 (a) Apple green (b) (i) Sulfuric acid (1) White precipitate (1) (ii) $Ba^{2+} + SO_4^{2-} \rightarrow BaSO_4$

Q3 (a) $Mg + H_2O \rightarrow MgO + H_2$

(b) No colour.

Q4 (a) To strengthen teeth / reduce tooth decay.

(b) Add 32×10^{-6} mol F^- per dm^3 (1); Add 5.33×10^{-6} mol Na_2SiF_6 (1); M_r (Na_2SiF_6) = 188.1; Add 0.010 g of Na_2SiF_6 (1).

Q5 (a) Si as this is the only giant covalent structure.

(b) Ar as ionisation energies increase across a period.

(c) Cl as electronegativity values increase across the period, up to group 7.

Q6 (a) mass of chlorine = $1.6 \times 10^6 \times 1000 \times 1000 = 1.6 \times 10^{12}$ g (1).
moles of chlorine, Cl_2, = 2.25×10^{10} mol (1).
Award 1 mark in total for 4.5×10^{10} (moles of Cl atoms).

(b) A *p*-block element has its outermost electron in a *p*-orbital.

(c) Kills bacteria / Disinfects.

(d) (i) $Cl_2 + 2Br^- \rightarrow 2Cl^- + Br_2$

(ii) Chlorine is a stronger oxidising agent than bromine (do not accept bromide) so it can oxidise bromide to bromine (1).

Iodine is a weaker oxidising agent than bromine so cannot oxidise bromide (1).

(e) (i) Lilac flame (do not accept any reference to white flame).

(ii) M_r $(KMgCl_3)$ = 169.9, M_r (H_2O) = 18.02 (1).
61.1 / 169.9 = 0.3596; 38.9 / 18.02 = 2.1587 (1).
x = 2.1587/0.3596 = 6 (1).

(iii) 2 marks for 3 correct observations, 1 mark for 2 correct observations:

Solution added	Observation
Aqueous sodium hydroxide	White precipitate
Dilute sulfuric acid	No visible change
Aqueous silver nitrate	White precipitate

Q7 Indicative content:

- Flame test for sodium is golden-yellow, potassium is lilac, barium is apple-green.
- Flame test for magnesium gives no colours.
- Test for halide using nitric acid followed by silver nitrate (MUST note nitric acid).
- Chloride gives white precipitate, bromide gives cream precipitate, iodide gives yellow precipitate.
- Addition of ammonia will cause bromide precipitate to dissolve but iodide does not.
- Carbonate will cause fizzing upon addition of nitric acid.
- Barium chloride can be used to test for sulfates, where it will give a white precipitate to differentiate between potassium nitrate and potassium sulfate.
- Magnesium nitrate can be used to test for hydroxides, where it will give a white precipitate to differentiate between sodium hydroxide and sodium nitrate.

0 marks: No relevant content
1–2 marks: At least 4 relevant observations from points 1–6.
3–4 marks: Method starts with flame tests and nitric acid/silver nitrate tests and uses these to correctly identity five compounds. At least 6 relevant observations from points 1–6.
5–6 marks: Method starts with flame tests and nitric acid/silver nitrate tests and uses these to correctly identity five compounds. Remaining pairs (sodium hydroxide/sodium nitrate and potassium sulfate/potassium nitrate) differentiated using appropriate tests. At least 8 relevant observations from points 1–6.

Q8 (a) Barium (Must give reason to gain this mark) (1).

Ionisation energies decrease down the group, so it is easier for barium to lose its outer electrons and react (1).

(b) Magnesium (Must give reason to gain this mark) (1).

Magnesium has more outer electrons than sodium (or numbers of outer electrons given) so stronger metallic bonding is formed (1).

(c) An oxide that will react with acids (to form a salt).

(d) (i) $MgCO_3 \rightarrow MgO + CO_2$

(ii) (I) Any temperature in the range 1000–2000 °C.

(II) $M_r (BaCO_3) = 197$, $M_r(BaO) = 153$ (1).
Percentage remaining = $(153/197) \times 100 = 77.7\%$ (1).

(e) Use standard/volumetric pipette in place of measuring cylinder (1). Add the acid in smaller amounts/drop by drop (1). This will make the measurements will be more precise (1).

Repeat the titration without indicator (1) using the same volume of acid (1). This will make the final salt pure / stop the salt being contaminated by indicator (1).

ANY five points.

Q9 (a) N or P   (b) Sr   (c) F/Cl/Br/I   (d) Mg;   (e) Cs   (f) Mg   (g) F

Q10 (a) Flame test (1)   Lilac flame (1).

(b) (i) Ca^{2+} (aq) + CO_3^{2-} (aq) $\rightarrow$ $CaCO_3$ (s).

(ii) To react with ALL the carbonate ions present.

(iii) M_r ($CaCO_3$) = 100.1 so 208.2 contains 108.1 ÷ 18.02 = 6 water molecules $CaCO_3.6H_2O$ (1).

M_r of first stage = 5.67/10 × 208.2 = 118.05 so $CaCO_3.H_2O$

M_r of second stage = 4.81/10 × 208.2 = 100.1 $CaCO_3$

M_r of final stage = 2.69/10 × 208.2 = 56.0 so CaO

(1 for any three M_r values or formulae, 2 marks for all six).

Heating causes water to be lost until the sample is anhydrous (1).

Further heating causes the calcium carbonate to decompose to calcium oxide (1).

(iv) Barium has a greater A_r OR barium carbonate has a greater M_r so precipitate would have a greater mass (1).

Larger mass would lead to a smaller percentage error (1).

Section 1.7: Simple equilibria and acid-base reactions

Q1   pH = $-$ log[H^+]

Q2   A reversible reaction where the rates of forward and reverse reactions are equal.

Q3 $K_c = \dfrac{[SO_3]^2}{[SO_2]^2\,[O_2]}$

Q4   0.70

Q5   Weak acid = an acid that only partially dissociates in water (1).
Dilute acid = small amount of acid dissolved in large volume of water (1).

Q6 (a) $K_c = \dfrac{[CH_3COOH]}{[CH_3OH]\,[CO]}$

(b) 4.47×10^{-5} mol dm^{-3}

Q7 (a) M_r (HI) = 128.01 so moles HI = 15.0 ÷ 128.01 = 0.117 mol (1).

Concentration = 0.117 ÷ 0.25 = 0.469 mol dm^{-3} (1).

pH = $-$log(0.469) = 0.33

(b) (i) $K_c = \dfrac{[HI]^2}{[H_2][I_2]}$

(ii) $[HI]^2$ = $K_c \times [H_2] \times [I_2]$ (1) = 19.0 (1) so [HI] = 4.36 mol dm^{-3} (1).

(iii) Equilibrium constant is greater, so amount of product is greater (1).

This means that the equilibrium has shifted to the right as the temperature increased (1). Therefore, the forward reaction must be endothermic (1).

(iv) No effect (1) as there is no change in the number of gas molecules in the reaction (1).

Q8 (a) $K_c = \dfrac{[NO_2]^2}{[N_2O_4]}$

(b) (i)   Mass NO_2 = 2.53 − 0.38 = 2.15 g (1)
Moles NO_2 = 2.15 ÷ 46 = 4.67×10^{-3} (1)
Concentration of NO_2 = 4.67×10^{-3} ÷ (100/1000) = 0.467 mol dm^{-3} (1)

(ii) Concentration of N_2O_4 = (0.38 ÷ 92) ÷ (100/1000) = 0.041 mol dm^{-3} (1)
K_c = 0.467^2 ÷ 0.041 = 5.29

(c) Increasing pressure will shift the equilibrium to the side with fewest gas molecules (1). This is to the left, reducing the amount of NO_2 present (1).

Q9 (a) Increasing temperature shifts the equilibrium in the endothermic direction (1). This is to the left and reduces yield of ammonia (1).

Increasing pressure shifts equilibrium in direction with fewest gas molecules (1). This is to the right and increases the yield of ammonia (1).

(b) (i) $2NH_3 (aq) + H_2SO_4 (aq) \rightarrow (NH_4)_2SO_4 (aq)$

(ii) Measure 25 cm³ of ammonia using a volumetric pipette (1). Add a few drops of indicator (1). Titrate using sulfuric acid until the endpoint is reached and there is a permanent colour change (1). Repeat the experiments, using the same volumes but no indicator (1). Evaporate majority to water and allow to cool and crystallise (1). Filter then heat crystals to 100 °C to reach constant mass (1). MAX 5 points.

Q10 (a) (i) $M_r(Na_2CO_3) = 106$ (1) Mass = 2.65 g (1).

(ii) • Dissolve sodium carbonate in 100 cm³ of water in a beaker (1).
  • Transfer to 250 cm³ volumetric flask using a funnel (1).
  • Wash beaker/funnel/glass rod three times with water going into volumetric flask (1).
  • Add water up to 250 cm³ line and shake to mix thoroughly (1).

(b) $Na_2CO_3 = 2.50 \times 10^{-3}$ mol (1); $HCl = 5.00 \times 10^{-3}$ mol (1); Conc. HCl = 0.292 mol dm⁻³ (1)

Q11 (a) 1 mark for table, 1 mark for both means.

	1	2	3	4
Volume that reacted with NaOH / cm³	16.05	15.40	15.55	15.45
Volume that reacted with NaHCO₃ / cm³	7.85	7.85	7.90	7.85

Mean volume that reacts with NaOH = 15.47 cm³

Mean volume that reacts with NaHCO₃ = 7.86 (allow 7.87) cm³

(b) (i) More reliable as values are greater so percentage errors are smaller.

(ii) More reliable as the values are closer together / all values are concordant.

(c) (i) Moles that react with NaOH = 1.73×10^{-3} mol so concentration = 6.91×10^{-2} mol dm⁻³
Moles that react with NaHCO₃ = 3.40×10^{-3} mol so concentration = 0.136 mol dm⁻³
1 mark for each moles, 1 mark for both concentrations.

(ii) Mass of NaHCO₃ in 25 cm³ = 0.286 g (1). Mass of NaOH in 25 cm³ = 6.92×10^{-2} g (1).
% NaHCO₃ by mass = [0.286 ÷ (0.286 + 0.0692)] × 100 = 80.5% (1).

Q12 (a) Increasing temperature shifts the equilibrium in the endothermic direction (1). This is to the right, giving a greater yield of ethyl ethanoate (1).

(b) (i) Strong means the acid is fully dissociated.

(ii) A dehydrating agent removes water from the reaction (1). This shifts the equilibrium to the right to produce more water, and hence more ethyl ethanoate (1).

(c) Adding concentrated sulfuric acid increases the concentration of H⁺ ions (1). This will cause the equilibrium to shift to the left, reducing the amount of dissociation of the ethanoic acid (1).

Q13 (a) Error in each measurement = 0.005 g so error in two measurements = 0.01 g (1).
Percentage error = (0.01 ÷ 5.74) × 100 = 0.17% (1).

(b) $CaCO_3 + 2HCl \rightarrow CaCl_2 + CO_2 + H_2O$

(c) 0.200 mol of HCl

(d) Moles NaOH = 18.20 × 0.500/1000 = 9.10×10^{-3} mol (1).
Moles HCl = 9.10×10^{-3} mol (1).

(e) Moles reacted HCl in 250 cm³ = $(0.020 - 9.10 \times 10^{-3}) \times 10 = 0.109$ mol (1).
Moles of CaCO₃ = 0.109 ÷ 2 = 0.0545 mol (1).
Mass CaCO₃ = 5.46 g (1).
Percentage by mass = (5.46 ÷ 5.74) × 100 = 95.0% (1).

Q14 (a) A weak acid dissociates partially/in a reversible reaction (1) to release/donate H⁺ (1).

(b) $CuO + 2CH_3COOH \rightarrow (CH_3COO)_2Cu + H_2O$

(c) (i) $[H^+] = 10^{-pH} = 1.995 \times 10^{-3}$ mol dm⁻³

(ii) Moles of NaOH = 5.525×10^{-3} mol
Concentration of acid in mol dm⁻³ = $5.525 \times 10^{-3} ÷ (25.0/1000) = 0.221$ mol dm⁻³ (1).
Concentration of acid in g dm⁻³ = 0.221 × 60.04 = 13.27 g dm⁻³ (1).

Practice questions Unit 2: Energy, Rate and Chemistry of Carbon Compounds

Section 2.1: Thermochemistry

Q1 Energy of products higher than the reactants (1).
Activation energy labelled (1).

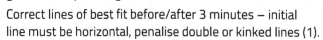

Q2 (a) Correct scale – temperature on y-axis and graph occupies over half of the available grid and does not start at zero on y-axis (1).

Correct plotting of points to +/– one small square – lose this mark if graph plot including extrapolations goes off the printed grid (1).

Correct lines of best fit before/after 3 minutes – initial line must be horizontal, penalise double or kinked lines (1).

Correct extrapolations to the third minute – both lines must be extrapolated to 3rd minute (1).

(b) Correct value for temperature rise should be linked to the graph.
Ideal answer is 27.0 – 20.3 = 6.7 °C (answer given to 1 dp) (1).

(c) $q = mc\Delta T$ (1).
= 50 × 4.18 × 6.7 = (ecf using answer to 2(b)) (1).
= 1400 J (1).

(d) (i) $NaOH + HCl \rightarrow NaCl + H_2O$.

(ii) Number of moles of NaOH (or HCl) = 1.00 × 25/1000 = 0.025 (1).
$\Delta H = -q/n = -1400$ (or answer from 2(c))/0.025 = 56012 (1).
= −56.0 kJ mol^{-1} (1).

Lose 1 mark if answer not negative.

(e) Answer should be same value as answer in 2(d)   (1).

Explanation: nitric acid is also monoprotic/reacts in 1:1 ratio like HCl with NaOH/ owtte (1).

Q3 (a) The enthalpy change when 1 mole of a substance is burned (1).
Completely (or in excess oxygen) under standard conditions (1).

(b) $C(s) + 0.5O_2(g) \rightarrow CO(g)$.

(c) One of the following:

CO is not the only product, e.g. CO_2 is also formed
Complete combustion occurs
Further oxidation occurs

(d) Correct energy cycle or $\Delta H_r = \Sigma \Delta H_f(\text{products}) - \Sigma \Delta H_f(\text{reactants})$ (1).
$\Delta H = 174 - [(5 × 393) + (6 × 286)]$ (1).
= −3507 kJ mol^{-1} (1).

Correct answer gains full marks

+3507 gains 1 mark only

ECF allowed if arithmetic error made.

(e) The two enthalpy changes are for the same reaction / have the same reactants and products (owtte).

Q4 (a) ΔH = Sum of bond enthalpies in reactants – Sum of bond enthalpies in products (1).
= [(5 × 412) + 348 + 360 + 463 + (3 × 496)] – [(4 × 743) + (6 × 463)] (1).
= 4719 – 5750 (1).
= −1031 kJ mol^{-1} (1).

Correct answer gains full marks
Max 2 marks if sign incorrect
ECF allowed if arithmetic error made.

(b) This is because average bond enthalpy values used are not exact values but are averaged from bond enthalpy values from a range of compounds.

Section 2.2: Rates of reaction

Q1   Basic diagram (1).

Labelled reactants and gas syringe (1).

Q2   (a) Moles of magnesium = 0.5/24 = 0.02 mol (1).

Moles of HCl = 2.0 × 0.025 = 0.05 mol (1).

0.02 mol Mg needs 0.04 mol HCl to react with, so acid is in excess.

(b) Graph:

x-axis label = time (s) and y-axis label = volume (cm^3) and sensible scales that mean graph covers over half the grid (1).

data points plotted +/− ½ small square lose one mark per mistake (2).

best fit line through points (1).

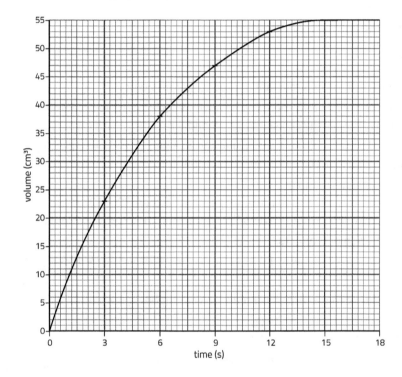

(c) Tangent to the curve drawn at time = 0 (1).

Gradient calculated e.g. dy/dx = 12/1.5 = 8 cm^3/s (1).

(d) Overall rate = total volume/time for reaction to end

e.g. 55/14 (time should be taken from graph but allow 15)

= 3.9 cm^3/s (1).

Overall rate is lower because it is the average rate for the whole reaction (1)

and the rate decreases with time (1).

(e) The magnesium particles have all been used up and so there are no more collisions/reactions (between acid and metal) (1).

Reaction has ended/no more gas is produced (1).

Q3

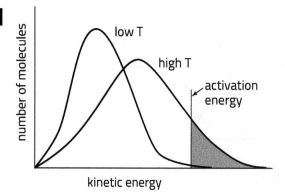

axes labelled:

y: number *(or fraction or %)* of molecules *(or particles)*   x: energy *(or KE)* (1).

curve starts at origin (1).

skewed to right (1).

approaches x-axis as an asymptote *(penalise a curve that crosses the energy axis)* (1).

second curve displaced to the left (and does not cross high T curve for a second time) (1).

and low T peak higher   (1).

Explanation: (could be annotated on graph)

many fewer molecules (1).

have $E > E_a$ (1).

Q4 (a) The change in concentration of a reactant or product per unit time.

(b) A lower concentration means that there will be fewer reactant particles in a fixed volume and (1) so there will be fewer successful collisions per second – resulting in a slower reaction (1).

(c) (i) By timing the appearance of the colour using e.g. a colorimeter either collecting results continuously / or timing a certain colour change (owtte).

(ii) Comparing experiments 1 and 2:
The concentration of A triples and rate changes by a factor $1.89 \times 10^{-3}/6.30 \times 10^{-4} = 3$ (1).
So there is a direct relationship between the concentration of A and the rate (1).

(iii) Yes the concentration of B affects the rate (1).
If we compare experiments 2 and 3 the rate stays the same but the concentration of A has doubled, so B must also have an effect (owtte) (1).

Q5 (a) A catalyst is a substance that speeds up a reaction without being used up.

(b) A catalyst provides an alternative route (1).
With a lower activation energy (1).

(c) A homogeneous catalyst is in the same state as the reactants.

(d) Concentrated sulfuric acid.

(e) Lower temperatures can be used and so energy costs are lower.

Section 2.3: The wider impact of chemistry

Q1   Answer should cover at least half of the following principles and should be illustrated by reference to each of the processes: (6)

- Waste prevention – only one product from hydration, fermentation will generate waste from plants used to obtain sugar, the actual fermentation process also creates waste.

- Best atom economy – hydration is 100%.

- Safest methods and chemicals – processes, such as fermentation, operate at near normal temperatures and pressures are safer, glucose is less flammable than ethene.

- Energy efficiency – fermentation does not require high temperatures or pressures and so has a lower energy demand.

- Use of renewable versus non-renewable feedstocks/raw materials – ethene from crude oil is non-renewable but widely available in oil-producing countries, whereas glucose is renewable and available from sugar cane grown in warm countries.

- Use of catalysts – both use catalysts but enzymes are far more efficient.
- Least environmental impact – the CO_2 given off in fermentation is taken in by plants as they grow/photosynthesise, so fermentation has the possibility of being carbon neutral.

Q2 Answer should include any three from the following:

- Renewable raw materials rather than non-renewable will ensure that the process is sustainable.
- An energy-efficient process that uses alternative energy sources such as wind power rather than a process which burns fossil fuels releasing carbon dioxide and possible causing acid rain.
- A process with high atom economy which means less waste, and if there is any waste it should be managed by biodegradation rather than using incinerators or landfill.
- A process that uses catalysts so that it can be operated at a lower temperature which reduces energy demands.
- A safe process that avoids the use of high temperatures and high pressures and avoids the use or production of toxic chemicals.

Plus:

- Some kind of reference to low impact upon the community from, e.g. lorries transporting materials, noise and dust pollution, release of harmful chemicals, etc., whilst creating lots of jobs and other positive aspects.

Section 2.4: Organic compounds

Q1 (a) Structural isomerism – when compounds with the same molecular formula (1) exist in different structural forms (1).

(b) Accept any type of structure (skeletal, etc.) but do not accept *E-Z* isomers as these are stereoisomers (4).

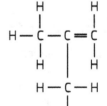

| | chlorocyclopropane | 2-chloropropene | 3-chloropropene |

Q2 (a)

(b) This time the *Z* isomer is an acceptable answer, or any structural isomer, all drawn in any format. Any one of:

but-1-ene   but-2-ene   2-methypropene

cyclobutane   methycyclopropane

(c) CH_2

Q3 (a) Mass of $CH_3O = 31$ and $62/31 = 2$ (1).
molecular formula = $C_2H_6O_2$ (1).

(b) There have to be 2 alcohol groups and so the answer could be the 1,1 or 1,2 isomer. Credit 1 mark for drawing one of these and award the second mark if all bonds are shown (2).

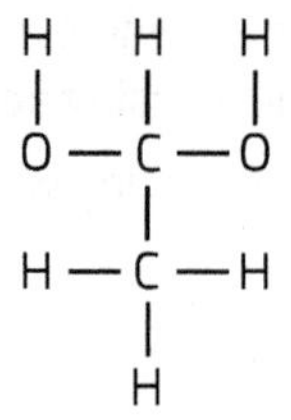

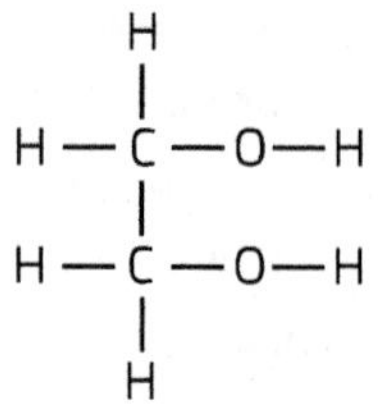

(c) Ethane-1,1-diol or ethane-1,2-diol (depending upon structure drawn) (1).

Q4 (a) (2-)Methylbutane, (2,2-)dimethylpropane, pentane (numbers not needed) (3).

(b) The less branching of the alkane chain the higher the boiling temperature (or vice versa) – owtte (1).

The greater the surface contact (the straight chain isomer) the greater the Van der Waals forces between the molecules, and so the greater the amount of energy needed to separate them (2).

Q5

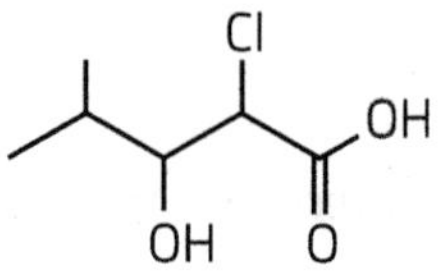

Q6 (a) C_3H_6O

(b)

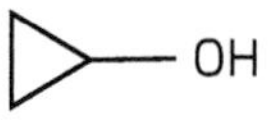

Q7 (a)

C	H	Cl	O	
18.60/12	1.57/1.01	55.03/35.5	24.80/16	
1.55	1.55	1.55	1.55	(2).

Empirical formula = CHClO (1).

(b) Observation with sodium hydrogencarbonate shows that a carboxylic acid group is present.

(c) Observation with sodium hydroxide shows that only one carboxylic acid group is present.

(d) Compound X = $CHCl_2CO_2H$ (1).
Reasoning:
There must be a COOH group present but only one (1).
The simplest multiple of the empirical formula that gives one carboxylic acid group is 2 × CHClO (1).

Section 2.5: Hydrocarbons

Q1 (a) Fractional distillation

(b)

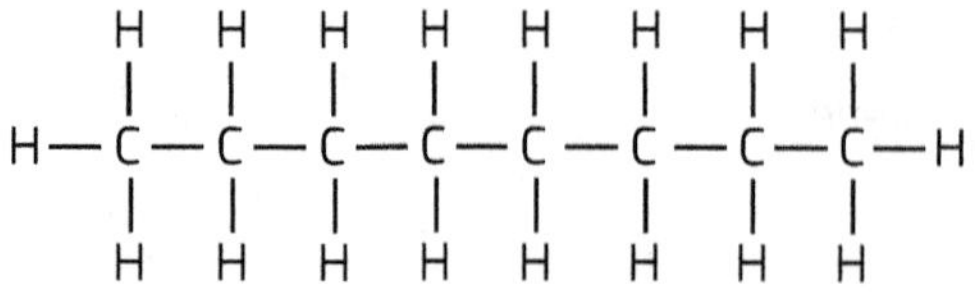

(c) 2,2,4-trimethylpentane

(d) Isooctane is more branched than octane and so the Van der Waals forces between the molecules are weaker (and so less energy is needed to separate the molecules) (2).

(e) $C_8H_{18} + 8.5O_2 \rightarrow 8CO + 9H_2O$
Correct formulae (1).
Balancing (1).

(f) Carbon monoxide is toxic.

Q2 (a)

C	H	Cl
37.8/12	6.3/1.01	55.9/35.5
3.15	6.24	1.57
2	3.97	1

Empirical formula = C_2H_4Cl (2).

(b) Mass of C_2H_4Cl = 63.54

127/63.54 = 2 so molecular formula = $C_4H_8Cl_2$

(c) Butane.

(d) (Free) radical substitution.

(e) Initiation: $Cl_2 \rightarrow 2Cl^{\bullet}$ (1).

Propagation: $CH_3CH_2CH_3 + Cl^{\bullet} \rightarrow CH_3CH_2CH_2^{\bullet} + HCl$ (1).

plus one mark for showing structure of propane. Do not award this mark if molecular formulae used throughout.

$CH_3CH_2CH_2^{\bullet} + Cl_2 \rightarrow CH_3CH_2CH_2Cl + Cl^{\bullet}$ (1).

Allow this termination reaction instead of the second propagation reaction:

$CH_3CH_2CH_2^{\bullet} + Cl^{\bullet} \rightarrow CH_3CH_2CH_2Cl$

(f) Further propagation reactions could occur, e.g. (1).

$CH_3CH_2CH_2Cl + Cl^{\bullet} \rightarrow CH_3CH_2CHCl^{\bullet} + HCl$

$CH_3CH_2CHCl^{\bullet} + Cl_2 \rightarrow CH_3CH_2CHCl_2 + Cl^{\bullet}$ (1).

Other termination reactions could occur: (1).

$2CH_3CH_2CH_2^{\bullet} \rightarrow CH_3CH_2CH_2CH_2CH_2CH_3$ (1).

Q3 (a) Answer should include:

- σ-bonds formed between C and H atoms and σ- and π-bonds formed between two C atoms (1).

- π-bonds formed by sideways overlap of p-orbitals (1).

- The π-bond restricts rotation about the double bond (1).

- Which means that when there are two different groups attached to the C atoms at the end of the double bond non-equivalent structures are possible (1).

- The E isomer is the isomer where the two highest priority groups attached to the C atoms at the ends of the double bond are diagonally distributed across the C=C (1).

(b) Accept any type of unambiguous structure. Isomers must be labelled (2).

  E isomer Z isomer

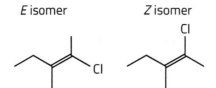

Q4 (a) Electrophilic addition.

(b)
dipole and curly arrow in HBr (1)
curly arrow from double bond (1)
curly arrow from lone pair on Br^- to carbocation (1)
where $R = CH_3CH_2CH_2$

(c) 2-bromopentane is formed from a secondary carbocation (1).

The alternative product would be formed from a primary carbocation which is less stable than a secondary carbocation (1).

(d) (i) Addition polymerisation.

(ii) (1 mark if draw polymer for but-2-ene) (2).

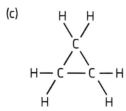

Q5 (a) PV = nRT so n = PV/RT (1).

$n = (1.01 \times 10^5 \times 36.5 \times 10^{-6})/(8.31 \times 373) = 1.19 \times 10^{-3}$ (1).

$M_r = 5.00 \times 10^{-2} / 1.19 \times 10^{-3} = 42.0$ (1).

Allow ecf

(b) $M_r = 42$ and the compound is a hydrocarbon.

3×12 (C) = 36 leaving 6 (1).

so molecular formula = C_3H_6 (1).

(c)

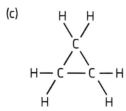

Q6 (a) The alkene must have 3 double bonds.

(b) (i) $CH_3CHCHCH_3 + H_2 \rightarrow CH_3CH_2CH_2CH_3$.

Do not accept molecular formulae.

(ii) The product (butane) will have the higher boiling temperature (1).

Both molecules have Van der Waals forces between them (1).

The Van der Waals forces between butane are stronger because there is greater surface contact between the molecules (owtte) (1).

(c) But-2-ene will also reacts with hydrogen bromide.

(i) $CH_3CHCHCH_3 + HBr \rightarrow CH_3CHBrCH_2CH_3$

Do not accept molecular formulae.

(ii) There is only one type of carbocation produced in this reaction (1)

and so each will be equally stable and products will form in equal quantities (1).

Or

But-2-ene is a symmetrical alkene (1)

and so it doesn't make a difference which way round the HBr adds (1).

Section 2.6: Halogenoalkanes

Q1 Nucleophile = species that can donate a lone pair of electrons to an electron-deficient species (1).

Substitution reaction = a reaction in which one atom/group is replaced by another atom/group (1).

Halogenoalkanes contain polar bonds: (1).

$^{\delta+}$ C–X $^{\delta-}$

Due to the halogen atom being more electronegative than the carbon atom, and so the electron-deficient carbon is attacked by the nucleophile causing the halogen atom to leave (and be replaced by the nucleophile) (1).

Q2 Indicative content (6):

CFCs are inert and accumulate in the upper atmosphere where UV radiation causes homolytic bond fission and the creation of chlorine radicals:

e.g. $CF_2Cl_2 \rightarrow CF_2Cl^{\bullet} + Cl^{\bullet}$ (any suitable example).

These chlorine radicals react with (and thus destroy) ozone:

e.g. $Cl^{\bullet} + O_3 \rightarrow ClO^{\bullet} + O_2$

These propagation reactions continue and lead to the formation of more chlorine radicals.

This forms a chain reaction – replenishing the number of chlorine radicals.

And so a small amount of CFC can lead to a large amount of ozone being destroyed.

This issue can be resolved if the compounds do not contain chlorine atoms (so chlorine radicals cannot be generated).

And so scientists have developed HFCs / halogenoalkanes with only fluorine in. C–F bonds are not broken to create fluorine radicals by UV light.

Q3 (a) e.g. 2-bromobutane.

(b) Sodium hydroxide and water (or aqueous ethanol) (2).

(c) e.g. $CH_3CH_2CHBrCH_3 + NaOH \rightarrow CH_3CH_2CH(OH)CH_3 + NaBr$.

(d) Suitable apparatus set up that 'works' (1).
Thermometer bulb opposite entry to condenser (1).
Water flow labelled correctly (1).

e.g.

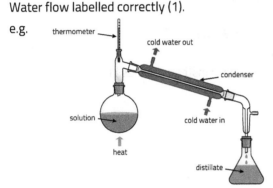

(e) It is likely that the student was comparing the boiling temperature of the product to the reactant (1).

The product is an alcohol with a higher boiling temperature than the corresponding halogenoalkane and so the product would not distil over leaving the reactant behind (1).

However, under these conditions (refluxing followed by distillation), the halogenoalkane would have been hydrolysed (1).

Therefore distillation would produce a mixture of butan-2-ol and water (which could be dried/ further separated) – owtte (1).

Q4 (a) Indicative content (6):

Firstly, a small amount of each would be placed in a test tube and some aqueous sodium hydroxide would be added.

The tube could be heated in a water bath for 5 minutes.
Then nitric acid would be added to neutralise any remaining sodium hydroxide.
Then a small quantity of silver nitrate solution is added and a precipitate forms.
e.g. $Ag^+(aq) + Cl^-(aq) \rightarrow AgCl(s)$

Silver chloride = white ppt; silver bromide = cream ppt; silver iodide = yellow ppt.
To confirm which precipitate is which, the final test is to add aqueous ammonia:
Silver chloride dissolves in dilute ammonia.
Silver bromide dissolves in concentrated ammonia.
Silver iodide does not dissolve.

(b) The order of hydrolysis is: Iodo > bromo > chloro (1).

The hydrolysis reaction is affected by the bond enthalpy of the carbon–halogen bond – the weaker the bond the faster the hydrolysis occurs (1).

The order of bond enthalpies is: C–I < C–Br < C–Cl (1).

(c) Order of hydrolysis: tertiary > secondary > primary (1).

Sensible reason
e.g. C–X bond weaker for tertiary halogenoalkanes (owtte) (1)
or tertiary halogenoalkanes could form stable carbocation intermediate and so bond breaks more easily, etc.

Q5 (a) Elimination.

(b) e.g. 3-chlorohexane or 3-bromohexane, etc.

(c) Sodium hydroxide and (dry) ethanol (2).

(d) e.g.
$CH_3CH_2CHBrCH_2CH_2CH_3 + NaOH \rightarrow CH_3CH_2CHCHCH_2CH_3 + H_2O + NaBr.$

(e) (i) Hex-2-ene could be produced as well from 3-bromohexane (1).
This occurs when elimination removes a hydrogen on the second carbon (as opposed to removing a hydrogen on the fourth carbon) (1).

(ii) Any form of unambiguous structure accepted (must be labelled) (2).

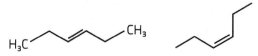

E-hex-3-ene   Z-hex-3-ene

(iii) Hexan-3-ol could be produced (1).
If the hydroxide ion behaved as a nucleophile instead of a base (1).

Q6 (a)

```
     H   F   H   H   H
     |   |   |   |   |
 I — C — C — C — C — C — H
     |   |   |   |   |
     H   F   F   H   H
```

(b)   OH
(structure)

(c) (i) Dichlorodifluoromethane.

(ii) Anything suitable e.g. can be easily evaporated, non-flammable, non-toxic.

(iii) Any fluorinated version of methane but not chlorinated. e.g.

(d) Solvent or anaesthetic.

Q7 (a) 1-chloroproapne has a more 'streamlined' structure (1).
That allows the molecules to pack closer together than the 2-chloropropane molecules (1).
And so the Van der Waals forces between 1-chloropropane molecules are stronger and more energy is needed to separate them (1).

(b) 1-chloropropane cannot form hydrogen bonds with water.

Section 2.7: Alcohols and carboxylic acids

Q1 Indicative content (6):

Reagents:
propan-1-ol
acidified potassium dichromate

Conditions:
Heat
And distil the product over as soon as it forms

Apparatus:

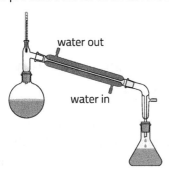

Equation:
$CH_3CH_2CH_2OH + [O] \rightarrow CH_3CH_2CHO + H_2O$
A primary alcohol must be used to produce an aldehyde.

It is necessary to distil the product as soon as it forms in order to prevent the aldehyde being further oxidised to propanoic acid.
$CH_3CH_2CHO + [O] \rightarrow CH_3CH_2COOH$

Q2 (a) Reactants (1).
Products (1).
Molecular formulae or ambiguous formulae (e.g. C_4H_9OH) not acceptable.
$CH_3CH_2CH_2CH_2OH + CH_3COOH \rightarrow CH_3COOCH_2CH_2CH_2CH_3 + H_2O$

(b) Concentrated sulfuric acid.

(c) Both the alcohol and the acid can form hydrogen bonds with water and are water soluble (1).

The ester, however, does not form hydrogen bonds and is immiscible (and less dense) and so floats on the surface of the water (1).

(d) Solvents, flavourings, fragrance.

Q3 (a) Hydration (allow addition).

(b) $C_2H_4 + H_2O \rightarrow CH_3CH_2OH$

(c) $300\,°C$, 60–70 atm, concentrated phosphoric acid catalyst (3).

(d) If crude oil was readily available and sunshine wasn't, countries might be more likely to make ethanol from ethene because they have a ready supply of ethene.

(e) Carbon neutral means that there is no net transfer of carbon dioxide to or from the atmosphere.

(f) Any two sensible suggestions of processes involved with the production of ethene (and ethanol) that release carbon dioxide, e.g.
Fractional distillation of crude, cracking, hydration, process, etc.
(and there are no processes that absorb carbon dioxide)   (2).

Q4 (a) Oxidation.

(b) Acidified potassium dichromate.

(c) There would be a colour change of orange to green.

(d) $CH_3CH_2CH_2CH(OH)CH_3 + [O] \rightarrow CH_3CH_2CH_2COCH_3$

Q5 Hydrogen bonding occurs between the ethylamine molecules (1).
Hydrogen bonding also occurs between the ethanol molecules (1).
Hydrogen bonding between ethanol molecules must be stronger and so more energy is needed to separate ethanol molecules as ethanol has a higher boiling temperature (1).

Q6 (a) $CH_3CH_2CH_2CH_2OH + 2[O] \rightarrow CH_3CH_2CH_2COOH + H_2O$

(b) Flask with vertical condenser (1).

    Unsealed apparatus (1).

    Heat (1).

This apparatus allows the mixture to be continuously heated without losing any volatile components because they condense and are returned to the reaction flask (owtte) (1).

(c) Propan-2-ol.

(d) Reactants correct (1).
Products correct (1).
$2CH_3CH_2CH_2COOH + Na_2CO_3 \rightarrow 2CH_3CH_2CH_2COONa + CO_2 + H_2O$

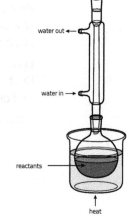

(e) Add excess (solid) sodium carbonate to pure butanoic acid (1).
Remove it by filtration (1).
Concentrate the solution by heating gently for a short while and then leave to allow crystallisation to complete. Collect crystals and dab with filter paper if damp (1).

Also acceptable would be the titration method using a solution of sodium carbonate. Once the titration is done with an indicator it must be repeated using the same volumes without an indicator. The solution would be converted to crystals in the same way as above.

Q7 (a) $CH_3CH_2CH_2OH + 4.5O_2 \rightarrow 3CO_2 + 4H_2O$
Correct formulae (1).
Correct balancing (1).

(b) $C_xH_yO_z + \text{excess } O_2 \rightarrow xCO_2 + 0.5yH_2O$
$x = 4.75/44 = 0.108 \qquad y = 2(2.43/18.02) = 0.270$ (1).
$z = (2 - [(0.108 \times 12) + (0.270 \times 1.01)])/16 = 0.027$ (1).
$C_{0.108}H_{0.270}O_{0.027}$ so divide through by 0.027 and get $C_4H_{10}O$ (1).

(c) (i) A fuel produced using a biological source.

(ii) Advantages include:
Renewable resource, does not require large amounts of energy to produce, can be carbon neutral under the right conditions (1).

Disadvantages include:
Raw materials only readily available in warm countries, large amounts of land required which could be used to produce food, large quantities of fertilisers (and water) needed to grow the crops (1).

Q8 (a) Butan-1-ol.

(b) Pentan-3-one.

(c) Correct compound – 3-chloro-2-methyl propanoic acid (1).
Correct structure (1).

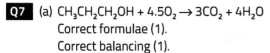

(d) Accept any tertiary alcohol with molecular formula of C_4H_9OCl (1).
Correct name (1).
e.g. $CH_3CH_2CCl(OH)CH_3$ 2-chlorobutan-2-ol
      $ClCH_2C(CH_3)(OH)CH_3$ 1-chloro-2-methylpropan-2-ol

Q9 (a) $C_6H_{12}O_6 \rightarrow 2CH_3CH_2OH + 2CO_2$

(b) A good answer will include photosynthesis, fermentation and distillation and may also include references to transportation and harvesting, e.g.
Carbon dioxide is absorbed during photosynthesis to produce the crops e.g. sugar cane (1).
Carbon dioxide is released during fermentation (1).

Carbon dioxide could be released during distillation if fossil fuels are used to provide heat (1).
As long as the total amount of carbon dioxide released during processing (including transportation)
does not exceed the amount of carbon dioxide absorbed during photosynthesis, then overall the
process will be carbon neutral (1).

Section 2.8: Instrumental analysis

Q1 Answer should cover these facts and the style should cover the phrase 'compare and contrast':

(a) Common to both types of NMR:

^{13}C NMR (2).
  - Tells us how many different carbon environments there are in a structure
  - and also the type of carbon environment (by using the chemical shift data in the data book).

^1H NMR (2).
  - Just like ^{13}C NMR, also tells us how many proton environments
  - and the type of environment (from the chemical shift value).

Found in (low resolution) ^1H NMR only: (1)
  - It is also possible to get more information from ^1H NMR spectra that is not available from the ^{13}C NMR – in ^1H NMR, comparing the relative peak heights gives us the ratio of how many hydrogens there are associated with each peak.

(b) Correct structures (2).

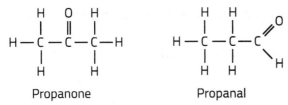

Propanone   Propanal

Answer should cover (most of) these facts:

^{13}C NMR (2).

Propanone

Would have two peaks due to two types of ^{13}C environments with two different chemical shift values – the two CH_3 carbons would give one peak 5 to 40ppm and the C=O carbon would give another peak 190 to 220ppm.

Propanal

Not having a symmetrical structure, would however have three peaks – one due to CH_2, one due to CH_3 both 5 to 40ppm and the third due to C=O 190 to 220ppm. This alone could distinguish between the two compounds but the correct structure could be confirmed by ^1H NMR.

^1H NMR (2).

Propanone

Propanone would have one large peak 2.0 to 2.5ppm due to 2 equivalent CH_3CO environments (only one peak because the structure is symmetrical).

Propanal

On the other hand, propanal would have 3 peaks. One peak about 1ppm (CH_3), one peak 2.0 to 3.0ppm ($-CH_2CO$) and a third peak about 9.8ppm (RCHO). The peak heights would be in the ratio 3:2:1 reflecting the number of hydrogens in each environment.

Q2 (a)

C	H	O
40.7/12	5.1/1.01	54.2/16
3.39	5.05	3.39
1	1.5	1

(1).

Empirical formula = $C_2H_3O_2$ (1).

(b) Mass of $C_2H_3O_2$ = 59.03
118/59.03 = 2 (1).
So molecular formula = $C_4H_6O_4$ (1).

(c) There are only two peaks but four C atoms in the molecule so the molecule must be symmetrical. (1).

Peak at ~180ppm corresponds to a COOH (1).

Peak at ~ 30ppm corresponds to a CH_2 (1).

NB. Insufficient to just copy from data booklet, answer needs to be specific to this molecule, so following would only gain 1 mark.

R — C — (carboxylic acid / ester)   160 to 185
        ‖
        O

— C — C —   5 to 40

And so the structure is (any unambiguous structure of butanedioic acid) (1).

Q3 (a) Peak furthest to the right-hand side gives m/z value for molecular mass (assuming z = 1).

(b) Fragmentation patterns tells us about the structure (1).

Ratios of certain peaks, e.g. molecular ion, can indicate isotopes, e.g. halogens such as chlorine and bromine (1).

(c) Structures (any unambiguous and labelled structures) (1).

Pentan-2-one   Pentan-3-one

Similarities:

Same molecular ion peak value (1).

Both would have similar fragments – example given (1).

e.g. m/z = 15 (CH_3^+) or m/z = 29 ($CH_3CH_2^+$) or any valid fragment – value and positive ion needed.

Differences:

Only pentan-2-one would have a fragment m/z = 43 (1).

This could be due to $CH_3CH_2CH_2^+$ or CH_3CO^+ (1).

Q4 (a) Mass of Cl = 100 – 24.7 – 2.1 = 73.2   (1).

C	H	Cl
24.7/12	2.1/1.01	73.2/35.5
2.06	2.08	2.06
1	1	1

(1).

Empirical formula = CHCl (1).

(b) (i) These peaks indicate the presence of isotopes (1).

Cl exists as ^{35}Cl and ^{37}Cl (1).

(ii) Largest mass of $CH^{37}Cl$ = 50 (50.01) (1).

150/50 = 3 so molecular formula = $C_3H_3Cl_3$ (1).

(c) Only one signal means there is only one H environment/symmetrical molecule (1).

Does not decolourise bromine water means C=C not present (not unsaturated) so structure must be cyclic to allow for so few H atoms in formula (1).

Y is (accept any unambiguous structure)   (1).

Q5 (a) Accept any unambiguous structure (3).

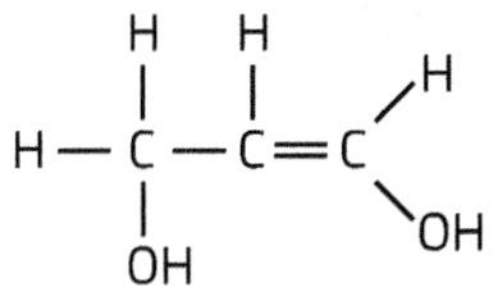

Propene-1,3-diol 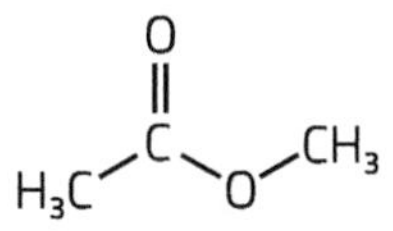   Methyl ethanoate 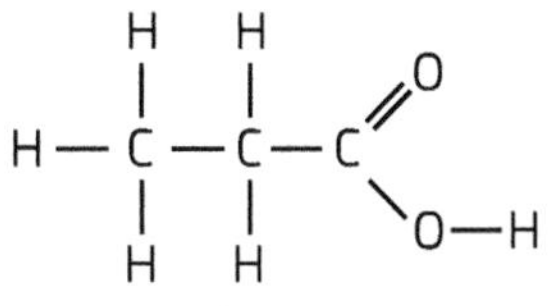   Propanoic acid

(b) Answers must quote IR values (as ranges from data book or appropriate values from spectra **and** assign these to appropriate functional groups):

A has a signal 3200 to 3550 indicating O–H (alcohol), there is no obvious signal at 1650 to 1750 cm^{-1} indicating absence of C=O group but there is a signal close to 1620 cm^{-1} which could indicate presence of C=C (1).

So A is propene-1,3-diol (1).

B has a signal 2500 to 3200 (very broad) indicating O–H (carboxylic acid) and there is also a signal at 1650 to 1750 cm^{-1} indicating presence of a C=O group (1).

So B is propanoic acid (1).

C is the only structure without an OH signal and it has an obvious signal 1650 to 1750 cm^{-1} indicating a C=O group (1).

So C is methyl ethanoate (1).

Unit 1 Practice paper answers

SECTION A

Q1 (a) $1s^2\ 2s^2\ 2p^6$

(b) $Na(g) \rightarrow Na^+(g) + e^-$ [Must include state symbols]

Q2 $pH = -\log[H^+]$

Q3 Metals have a sea of delocalised electrons that move to carry a current (1). Magnesium has two outer electrons that form the sea of delocalised electrons, aluminium has three. So with more delocalised electrons it is a better conductor (1).

Q4

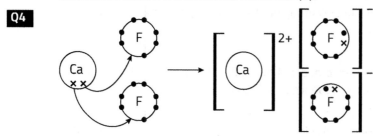

Q5 (a) $n = 3$ to $n = 2$

(b) $E = hf = 3.02 \times 10^{-19}$ J (1) 182.4 kJ mol^{-1} (1).

SECTION B

Q6 (a) NH_3 (1) as it can form hydrogen bonds with water molecules and the other two can't (1).

(b) PH_3 as it only has Van der Waals forces between its molecules (1). NH_3 can also form hydrogen bonds between its molecules that are stronger than Van der Waals forces (so NH_3 has a higher boiling temperature) (1). AsH_3 has more electrons than PH_3 so it forms stronger Van der Waals forces between molecules (so AsH_3 has a higher boiling temperature).

(c) A coordinate bond is a pair of shared electrons, both coming from the same atom (1). In a covalent bond the electrons come from different atoms (1).

(d) Moles $H_3PO_4 = 0.503$ mol 84.2% yield moles $PH_3 = 0.106$ mol (1) Mass $PH_3 = 3.60$ g (1).

Q7 (a) $A_r = (0.2783 \times 87) + (0.7217 \times 85)$ (1) $= 85.56$ (1).

(b) (i) Labels Rb^+ and Cl^- (1). Each ion has 6 ions arranged octahedrally OR cube with alternating ions (1).

(ii) Cs^+ ion is larger so can fit more Cl^- around it than Rb^+ can.

(c) (i) Outer electrons in same shell so same shielding/distance from nucleus (1) Sr has more protons in nucleus than Cs therefore greater nuclear charge and greater attraction electrons / electrons held more tightly (1).

(ii) Rubidium has more protons than potassium, but also more shielding (as outer electron is in different shell) (1). The greater shielding overcomes the greater nuclear charge so the effective nuclear charge in Rb is lower than Cs so outer electron of Rb is held less tightly (1).

(d) (i) Electron capture: ^{84}Kr; β^+ emission: ^{84}Kr; γ-emission: ^{84}Rb; β^- emission: ^{84}Sr (4 correct = 2 marks, 2 correct = 1 mark).

(ii) He^{2+} ion / helium nucleus / 2 protons and 2 neutrons.

Q8 (a) (i) $K_c = \dfrac{[NOCl]^2}{[NO]^2[Cl_2]}$

(ii) $[NOCl]^2 = 0.015^2 \times 0.012 \times 65000 = 0.1755$ (1) $[NOCl] = 0.419$ mol dm^{-3} (1).

(iii) Increasing temperature will shift the equilibrium in the endothermic direction (1). This is to the left, reducing product and reducing the value of K_c (1).

(b) The student is incorrect (1). NOCl has a lone pair in addition to bonds to Cl and O (1). Shape will be V-shaped / based on trigonal planar (1).

(c) Covalent bonds because difference in electronegativity is small (1). Polar N=O bonds due to difference in electronegativity, N−Cl non-polar as there is no difference in electronegativity (1).

(d) Strong acids dissociate completely (1) but weak acids only dissociate partially/ dissociate in a reversible reaction (1).

Q9 (a)

	1	2	3	4
Volume of acid for reaction in step 1 / cm³	15.50	15.45	15.55	15.50
Volume of acid for reaction in step 2 / cm³	27.35	26.90	26.95	27.00

Mean volume in step 1 = 15.50 cm³ (1). Mean volume in step 2 = 26.95 cm³ (1).
Penalise 1 mark if any value recorded to fewer than 2 decimal places.

(b) Step 1 as all values are close together / Step 2 as larger values have smaller percentage error.

(c) (i) Lower concentration would need a volume above 50 cm³ than wouldn't fit in burette.

   (ii) Higher concentration would reduce titration volumes so increase percentage error.

(d) (i) Moles of HCl = $15.50 \times 0.120/1000 = 1.86 \times 10^{-3}$ mol (1.) Moles $Na_2CO_3 = 1.86 \times 10^{-3}$ (1).
   Concentration of $Na_2CO_3 = 1.86 \times 10^{-3} \div 0.025 = 0.0744$ mol dm⁻³ (1).

   (ii) Moles of HCl = $26.95 \times 0.120/1000 = 3.234 \times 10^{-3}$ mol (1).
   Moles $NaHCO_3 = (3.234 \times 10^{-3}) - (1.86 \times 10^{-3}) = 1.374 \times 10^{-3}$ mol (1).
   Concentration $NaHCO_3 = 0.0550$ mol dm⁻³ (1).

Q10 (a) Indicative content:

TESTS:

- Flame test – calcium compounds give brick-red flame / Sodium compounds give golden-yellow flame / magnesium compounds give no colour.
- Solubility – magnesium carbonate, calcium carbonate and magnesium hydroxide are insoluble.
- Addition of acid – magnesium carbonate and calcium carbonate fizz and dissolve, magnesium hydroxide.
- Add silver nitrate solution to solutions of compounds – white precipitate due to chloride, cream precipitate due to bromide, yellow precipitate to iodide.
- Add concentrated ammonia solution to precipitates – silver bromide dissolves, silver iodide does not.

STRUCTURE:

- Only add acid to compounds that are insoluble.
- Start with either flame test or attempt to dissolve and only perform tests needed to distinguish pairs/threes of compounds.

5/6 marks: Plan that will identify all compounds. Appropriate structure that is efficient and does not perform additional tests. Bromide and iodide distinguished using ammonia.

3/4 marks: Plan that will identify all or almost all compounds. Method is not fully efficient and contains some additional tests.

1/2 marks: Majority of compounds identified correctly.

(b) Temperature between 200 and 750 °C as group 2 carbonates become more thermally stable down the group.

(c) (i) $Cl_2 + 2Br^- \rightarrow 2Cl^- + Br_2$

   (ii) Halogens become weaker oxidising agents down the group / halides become weaker reducing agents down the group (1). So chlorine Is a strong enough oxidising agent to oxidise bromide but iodine is a weak oxidising agent and cannot oxidise bromide (1).

Q11 (a) 4.608 g

(b) Percentage error = $100 \times (2 \times 0.0005) \div 4.608 = 0.022\%$

(c) mass H_2O = 1.514 g (1) $d = 6$ (1).

(d) moles $CO_2 = 314 \div 22.4 = 0.0140$ mol (1).
   moles $CO_3^{2-} = 0.0140$ mol (1)
   $b = 1$ (1).

(e) mass CuO = 2.226 g (1)
moles CuO = 0.028 mol (1)
$a = 2$ (1).

(f) M_r of georgeite = 4.608/0.140 = 329.14 (1).
Mass due to hydroxide = 34.02 (1).
Formula of georgeite = $Cu_2(CO_3)(OH)_2.6H_2O$ (1).

(g) Total negative ions = −2 + 2 × (−1) = −4 (1)   2 Cu = +4, so each Cu = +2 (1).

Unit 2 Practice paper answers

SECTION A

Q1 Either butan-1-ol or methylpropan-1-ol.

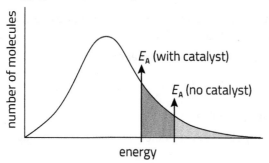

Q2 $C_8H_9NO_2$

Q3 Activation energy labelled on graph (with and without catalyst) (1).
So there are more successful collisions per second with a catalyst because more molecules have energy greater than or equal to the activation energy (owtte) (1).

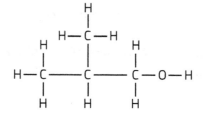

Q4 (a) Oxidation.

(b) Number of moles of ethanoic acid = 1 × 14/1000 = 0.014
Number of moles of sodium carbonate = ½ × 14/1000 = 0.007
Therefore volume of sodium carbonate = = 0.007/0.5 = 0.014 dm³ = 14.0 cm³.

(c) $CH_3COOH + CH_3CH(OH)CH_3 \rightarrow CH_3COOCH(CH_3)_2 + H_2O$.

(d) atom economy = (102/120) × 100 = 85%.

Q5 *E*-pent-3-en-2-ol (2).
1 mark if *E-Z* isomer not given

SECTION B

Q6 (a) Apparatus only (1).

Detailed labelling up to 2 marks   (2).

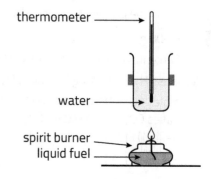

(b) $q = mc\Delta T$ (mark for correct formula or numbers in calculation) (1).

$= 6688$ J [$= 6.69$ kJ] (1).

0.46 g is 0.01 mol

So $\Delta H = -6.69/0.01 = -669$ kJ mol^{-1} (1).

(c) (i) Incomplete combustion.

(ii) Allow answers in range 2655 to 2680 kJ mol^{-1}.

(iii) Each alcohol increases by a CH_2 unit and so one more C–C and two more C–H must be broken (1).
Creating one more CO_2 and one more H_2O (owtte) (1).

Q7 (a) Values for rate.

New concentration / mol	Time / s	Rate / s^{-1}
0.010	32.5	0.031
0.008	39.1	0.026
0.006	55.0	0.018
0.004	85.9	0.012
0.002	129.0	0.008

(b) Labels for axes and more than half of grid used (1).
Points plotted correctly to +/– ½ square (1).
Straight line of best fit through middle range of points (1).

(c) Rate is proportional to concentration of hydrogen peroxide because graph gives a straight line.

(d) More particles of H_2O_2 in the same volume (1).
So there are more successful collisions per second (owtte) (1).

Q8 (a) (i) Nucleophilic substitution (1).

1 mark for lone pair and minus sign
1 mark for 2 curly arrows

(ii) One from the following:
- The yield is low
- Elimination occurs (ethene formed)
- The rate is slow
- Bromoethane has to be manufactured first / is expensive

(iii) Reaction 1: Elimination/dehydration (1).
Conc. phosphoric acid or conc. sulfuric acid (1).

Reaction 2: Electrophilic addition (1).
Hydrogen bromide (1).

(b) (i) Hydration / electrophilic addition.

(ii) Catalyst: (conc.) phosphoric (1).

Any two from: (2).
Pressure – 60–70 atmospheres
Temperature – 300 °C
Excess ethene/steam or recycle ethene or remove ethanol

Q9 (a) (i) (Free) radical substitution (1).
UV light (1).

(ii) Initiation $Cl_2 \rightarrow 2Cl^{\bullet}$ (1).
First propagation $CH_3CH_3 + Cl^{\bullet} \rightarrow CH_3CH_2^{\bullet} + HCl$ (1).
Second propagation $CH_3CH_2^{\bullet} + Cl_2 \rightarrow CH_3CH_2Cl + Cl^{\bullet}$ (1).
Termination $^{\bullet}CH_2CH_2Cl + Cl^{\bullet} \rightarrow CH_2ClCH_2Cl$ (1).
(Cannot use molecular formulae for last equation – structure must be clear.)

(b) (i) Both molecules have Van der Waals forces between them (1).
1,2-dichloroethane molecules are bigger with more electrons (1).
And so the Van der Waals forces between them are stronger and more energy is needed to separate them (1).

(ii) PV = nRT so V = nRT/P (1).
$n = 3.75/64.5 = 5.81 \times 10^{-2}$; $P = 2.02 \times 10^5$ Pa; T = 373 K (1).
$v = 8.92 \times 10^{-4}$ m^3 (1).
= 892 cm^3 (1).

(iii) number of moles = 5.81×10^{-2} so mass = $5.81 \times 10^{-2} \times 99 = 5.75$ g (1).
Volume = mass/density = 4.60 cm^3 (1).

Q10 (a) The enthalpy change for a reaction is independent of the route taken.

(b)
$$CH_3CH_2CH{=}CH_2 \ + \ HBr \rightarrow CH_3CH_2CHBrCH_3$$

$-[\Delta_fH(\text{but-1-ene})$
$+ \Delta_fH(HBr)]$ (1)

$\Delta_fH(\text{2-bromobutane})$ (1)

$4C(s) + 4\tfrac{1}{2}H_2(g) + \tfrac{1}{2}Br_2(l)$ (1)

Allow first arrow in upwards direction if negative sign removed.
State symbols not required for original equation/ignore.

(c) (i) Enthalpy change to break a bond (1).
Averaged over different molecules/environments (1).

(ii) Bond breaking = $(8 \times 412) + (2 \times 348) + 612 + 366 = 4970$ (1).
Bond forming = $(9 \times 412) + (3 \times 348) + 276 = 5028$ (1).

Enthalpy change for reaction = $4970 - 5028 = -58$ kJ mol^{-1} (1).
Marks allowed for ecf.

(d) The enthalpy values used in (c) were mean values not exact values and so the answer would not be as precise as the answer in (b).

Q11 (a) Compounds with the same molecular formulae but with different structures.

(b) (i) Reagents: Acidified potassium dichromate (1).
Observations: Colour change from orange to green with M and no colour change with N (2).

(ii) Reagent such as sodium carbonate or magnesium. Named indicator is not strictly a reaction but would probably be accepted (1).
Observations: No reaction with P but fizzing with Q (carbon dioxide or hydrogen) (2).

(iii) Reagent: bromine water (1).
Observations: Bromine water would be decolourised with R but there would be no change with S (2).

(c) (i) R would have a peak in the range 1620 to 1670 cm^{-1} due to C=C (1).
S, however, would not have a peak here as it has no C=C (1).

(ii) R would have 3 peaks due to 2 × CH on double bond, 2 × CH$_2$ next to it and 2 × CH$_2$ on opposite side of the ring. These peak heights/areas would be in the ratio 1:2:2 (2).
S would have only 1 peak as all 12H are equivalent so the peak would be intense/area under peak equivalent to 12H (2).